Anatomy For Holistic Practitioners
by
Colin Paddon Ph.D., D.Ac., D.N.M.

Anatomy For Holistic Practitioners
by
Colin Paddon Ph.D., D.Ac., D.N.M.

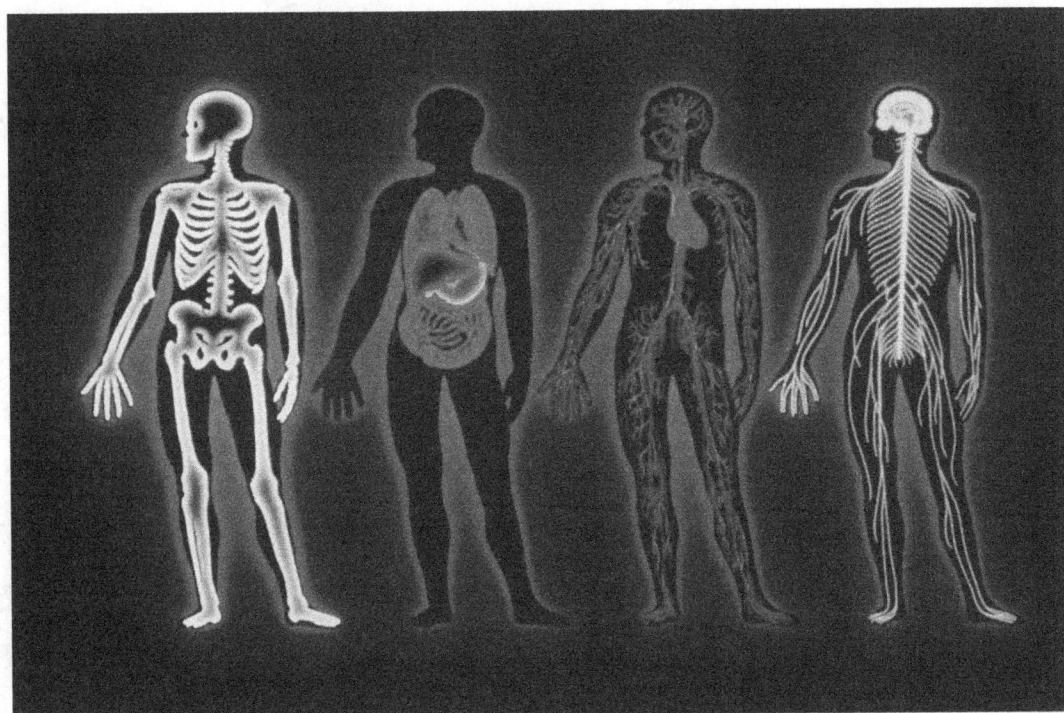

Airmid Holistic Books, Los Angeles

Airmid Holistic Books, Sunland, CA 91040
© Colin Paddon 2008. All rights reserved.
Published 2008.
Printed in the United States of America

ISBN 978-0-9796168-2-2

Contents

Bibliography

Principles of Anatomy & Physiology by Gerard Tortora & Nicholas Anagnostakos
Trail Guide to the Body by Andrew Beil
Atlas of Human Anatomy by Frank Netter
Grants Atlas of Anatomy by Anne Agur
Grant's Method of Anatomy by John Basmajiasn & Charles Slonecker
Nursing RGN by David Parkin
A.D.A.M. Student Atlas of Anatomy By Todd Olson
Anatomy & Physiology by Ethel Sloane
Anatomy Coloring Book by Wynn Kapit & Lawrence Elson
Human Anatomy & Physiology by Elaine Marieb
Anatomy and Physiology

Introduction

The following notes are subject matter relating to the amount of knowledge needed to become a therapist. Should you feel that further study would prove beneficial, then by all means feel free to do so. However, the course level of knowledge examination will be taken from material provided.

This course makes no pretence at enveloping the whole subject of anatomy and physiology, but is aimed at giving you the student a basic understanding of related anatomical parts, position and function.

Anatomy is defined as the study of the structure of the body. Physiology is defined as being the study of the function of those parts.

For example:
- The heart weighs approx. nine ounces, is pear shaped and lies on the left-hand side of the chest area and one third to the right. That is the basic anatomy.
- The heart pumps blood around the body after first oxygenating it via the lungs. That is the basic physiology.

By combining the knowledge of anatomy and physiology, we are able to get a clearer picture of the organ and its relationship to the body, together with its form and composition, and so, improve our understanding of the body as a whole.

For simplicity, the body will be divided into eight main systems. Some text books divide or sub- divide into more or less systems, however, for the purpose of this course we shall look at them as follows.

1.	The Skeletal System:	Bones, structure, support, articulation and mobility.
2.	The Muscular System:	Two types of muscle, voluntary and involuntary muscles and ligaments.
3.	The Vascular System:	Including the lymphatic system, heart and blood vessels.
4.	The Neurological System:	Covers most of the nerves of the body and brain.
5.	The Digestive System:	Including related organs of digestion.
6.	The Respiratory System:	Lungs.
7.	Genito-Urinary System:	Including the reproductive and kidney system.
8.	The Endocrine System:	The glands and their effects on the body.

All professional therapists should have a basic knowledge of medical terminology, which you will need to become familiar with to assist in understanding other professionals. Although we can not teach all the terms in this short course, those which you will use the most will be taught. Understanding this terminology will make subsequent study that much easier.

Historical

It is difficult to trace the exact study of anatomy and physiology, although the Egyptians with their famous embalming process must have gained a lot of knowledge about the human body as they perfected this art. It is interesting to note that the "℞" which European doctors write at the top of prescriptions, is in fact the "℞" symbol of the Eye of Horus, (the hawk headed sun god), who lost his eye in the battle and had it restored by Thoth, who was adopted as the patron God of physicians. Thoth was one of the many Gods invoked by ancient Egyptian doctors while administering their remedies.

It was with the Greeks that we learned in detail about anatomy and physiology. The first detailed accounts from this era came from Hippocrites, often referred to as the Father of Medicine.

Aristotle, is accredited as being the founder of comparative anatomy.

From the Roman era, a vast collection of surgical and dissecting instruments have been unearthed, which indicate a considerable understanding of anatomical form and function.

It was the second century A.D. that Galen lived, and his name is still remembered as being one of the greatest physicians and anatomists of antiquity. His work established the foundation of European anatomy as we know it.

From the sixteenth century medical school in Italy, Paracelsus Von Hohenheim, graduated to become a progressive medical teacher and did much to alter the accepted ideas of his day.

In 1543, Vesalius published his first drawings of the structure of the human body in his book "Fabric of the Human Body", and so paved the way for modern anatomy.

Since those early times many talented doctors have contributed to expand the knowledge of anatomy in the search to uncover and simplify the complex functions of the human body.

William Harvey is linked to the role and function of the heart and the process of oxygenated blood through the lungs.

Malpighi discovered capillary circulation in 1661.

Avenbrugger of Austria discovered "Percussion" In the middle of the 18th century.

Rene Laennec invented the stethoscope.

1822 Dr William Beaumont contributed much to the understanding of the function of the digestive
system.

1867 Lister discovered the antiseptic principles.

1877 Pasteur demonstrated the role of germs and disease.

1895 Roentgen discovered X-ray.

1904 Baylis and Stanley identified the first hormone.

1912 Frederick Gowland Hopkins, discovered vitamins.

1928 Alexander Fleming discovered antibiotics (penicillin)

1953 James Watson and Francis Crick discovered the double helix of DNA (Dioxyri-bonucleicacid).

Anatomy and physiology are subjects of continuous study and discovery.   Each year will bring new discoveries and understanding.

The Skeletal System

Skull

Clavicle

Scapula

Carpals

Humerus

Sternum

Ulna

Pelvis

Radius

Metacarpals

Femur

Phalanges

Patella

Tibia

Fibula

Tarsals

Metatarsals

Phalanges

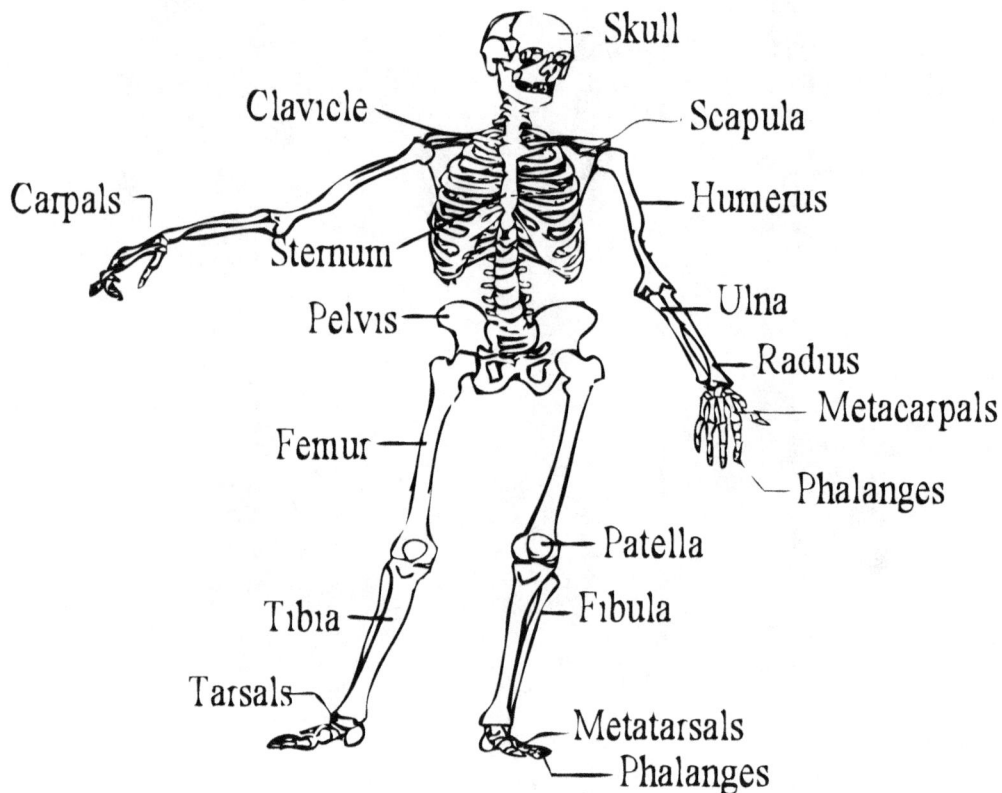

Principal Functions of the Skeletal System

The skeletal system provides the framework for the body, to give it shape, support and locomotion.

The two principal functions of the skeletal system are:

    1/ PROTECTION e.g.   The skull protects the brain
                             The rib cage protects the heart and lungs
                             The spinal column protects the spinal cord

    2/ LOCOMOTION e.g.   Movement.

Classification of Bones

208 BONES FORM THE SKELETON

Bone is dry, dense tissue composed of approximately:

Bone tissue is composed of

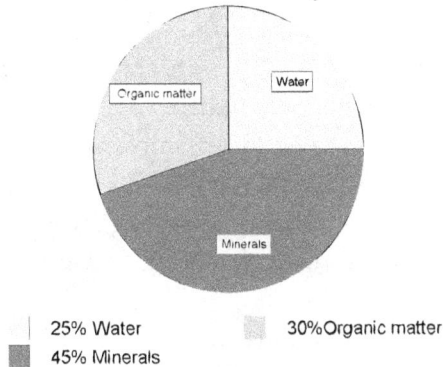

25% Water   30% Organic matter
45% Minerals

These figures are approximate and should not be taken as literal facts.

The mineral consists mainly of calcium phosphate, and a small amount of magnesium salts. These give the bone its rigidity and hardness.

The organic matter consists of fibrous material which gives the bone its toughness and resilience.

There Are Five Classifications of Bones:

1) **LONG BONES**:   Like the femur.
2) **SHORT BONES**:   Like the metatarsal bones.
3) **FLAT BONES**:   Like the frontal bone found in the skull.
4) **IRREGULAR BONES**:   Like the vertebrae.
5) **SESAMOID BONES**:   These are rounded masses found in certain tendons, the best example is the kneecap (patella)

In addition to the two principal functions of the skeletal system, the individual bones serve other purposes, such as the attachment for tendons and muscles as well as the formation of red blood cells in the bone marrow.

To provide locomotion or movement the skeleton provides us with a number of joints or articulations.
These fall into four main categories:

1. BALL AND SOCKET: Found in the hip and shoulder.
2. PIVOT: Found in the radius / ulna and the axis joint.
3. GLIDING: Found in the carpal and tarsal joints.
4. HINGE: Found in the knee, and a partial hinge is found in the elbow joint.

Besides the four main joint articulations, there are, of course, two other types of joint:

SYNARTHROSES — These are found in flat bones of the skull and are the fibrous joints that are thought to be fixed with no movement at all. (However, modern medicine has since disproved this by examining live samples, and found movement.)

AMPHIARTHROSES — These are found in joints like the spine which is a cartilaginous joint with slight movement. An example of this joint is found in the spine, where the discs of fibro-cartilage separate the bones of the spine. These moveable joints are supplied with a secretion called "Synovial Fluid". This is a whitish fluid, similar in consistency to raw egg white, which acts as a lubricant between the articulating surfaces.

The mucous bursae are sacs containing a clear viscous fluid which acts rather like a water cushion between joints. If the synovial membrane becomes inflamed, this is known as "synovitis" ("Tennis Elbow" a well known European terminology). If the bursae becomes inflamed this is known as "bursitis", (a well known example being "Housemaids Knee").

22 BONES FORM THE SKULL:

For the purpose of this course the principal bones to remember are:

- Parietal bone
- Temporal bone
- Sphenoid bone
- Zygomatic bone
- Frontal
- Maxilla
- Mandible
- Mental Foremen
- Infra-Orbital Foremen
- Occipital

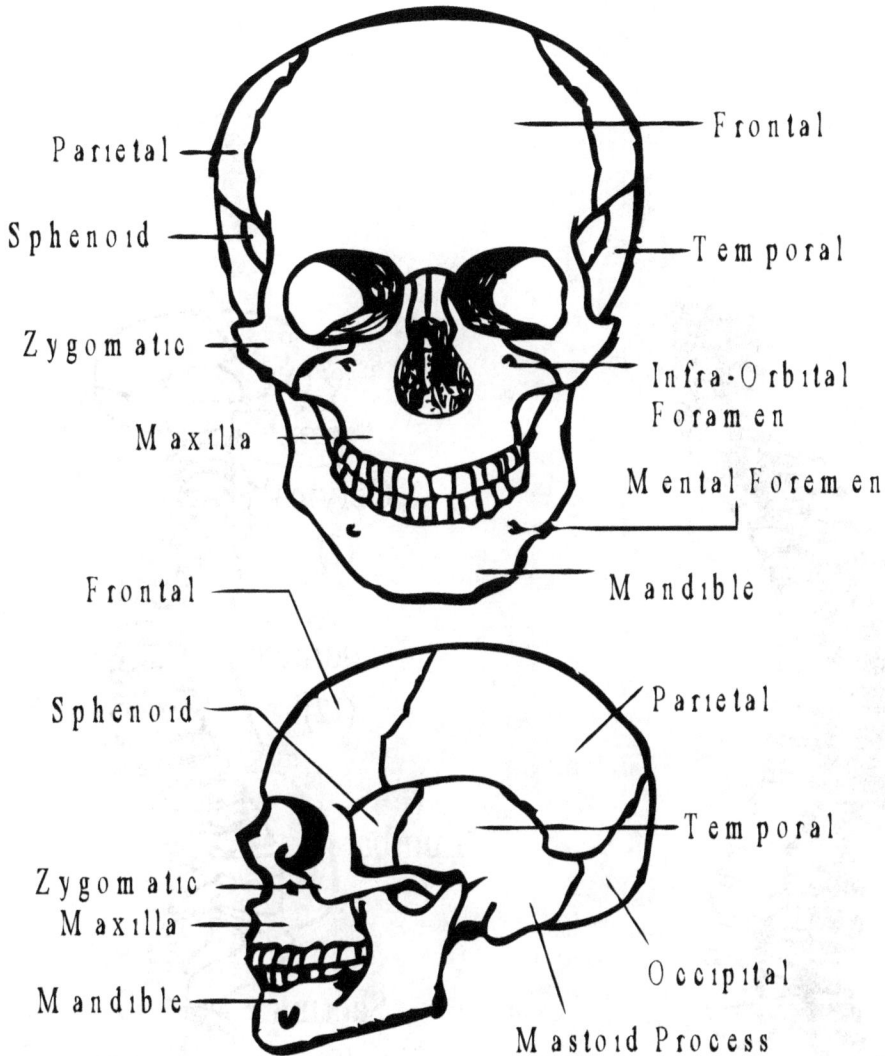

33 BONES FORM THE VERTEBRAE:

- Cervical vertebrae x 7
- Thoracic vertebrae x 12
- Lumber vertebrae x 5
- Sacrum (fused) x 5
- Coccyx (fused) x 4

Separating each vertebrae is a disc of fibro-cartilage. The top two cervical vertebrae are known as the Atlas bone and the Axis bone.

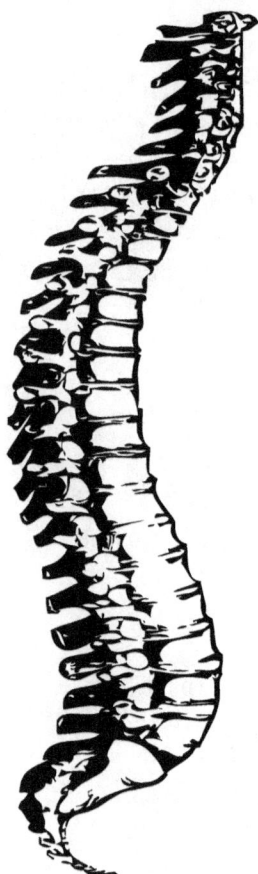

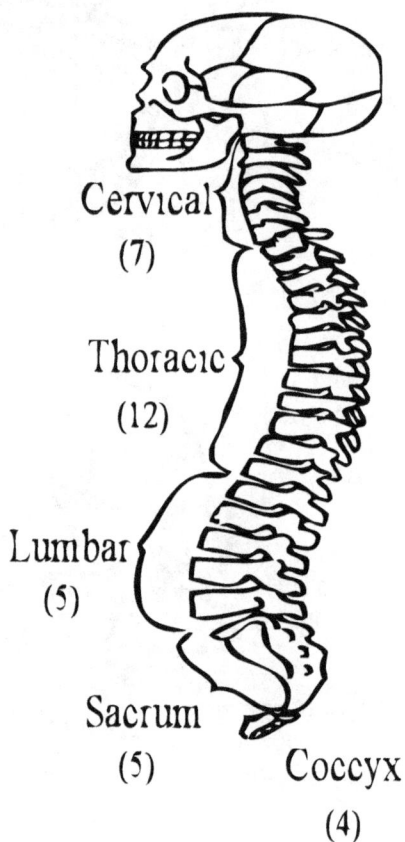

Cervical
(7)

Thoracic
(12)

Lumbar
(5)

Sacrum
(5)

Coccyx
(4)

25 BONES FORM THE THORAX (CHEST):

- The sternum, ribs x 12, (7 pairs of true ribs and five pairs of false ribs).
- The last two pair of false ribs are known as "floating ribs", because they are attached at the back, but not at the front. (They are not shown on this diagram).

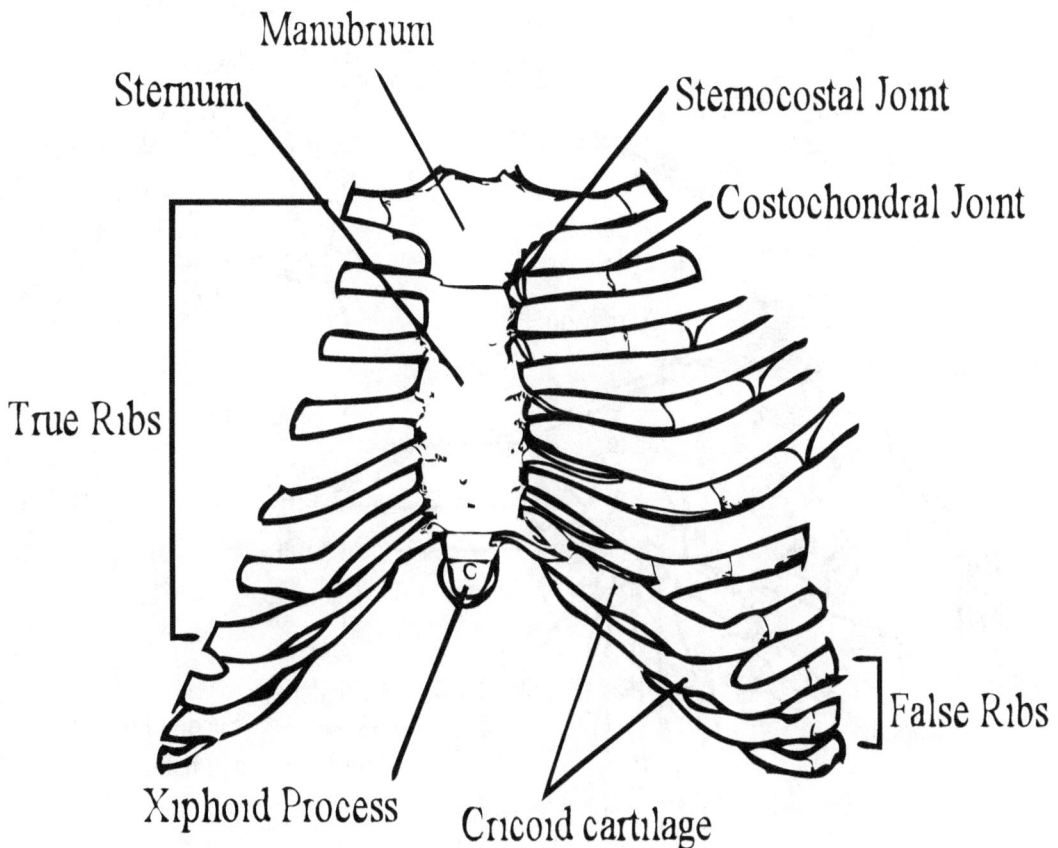

64 BONES FORM THE UPPER LIMBS OF BOTH ARMS (32 in each arm):

- Scapula
- Clavicle
- Humerus
- Radius
- Ulna
- Carpal bones x 8
- Metacarpal bones x 5
- Phalanges x 14

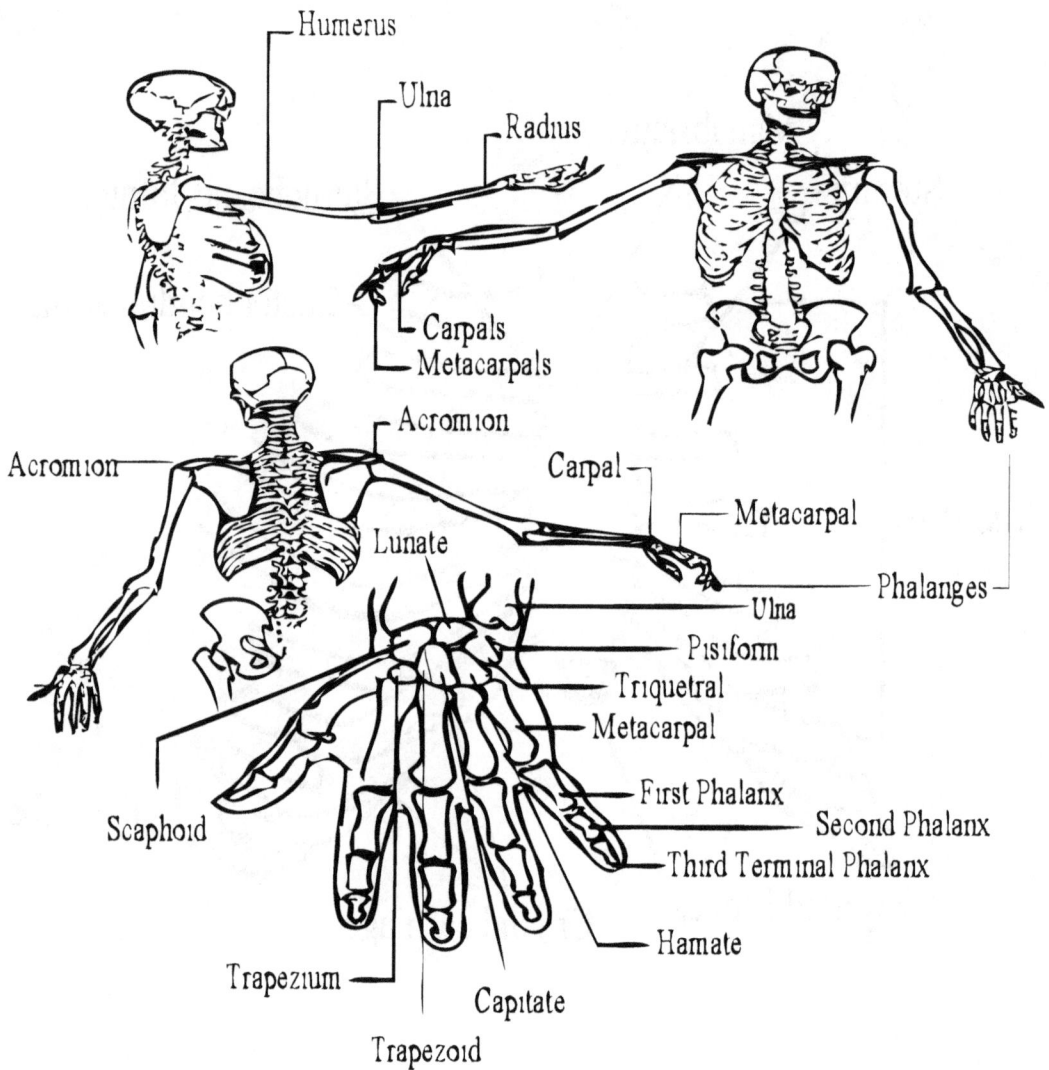

4 BONES FORM THE PELVIS:

Each innominate bone consists of the Ilium and Pubis.

- Right innominate bone
- Left innominate bone
- Sacrum
- Coccyx

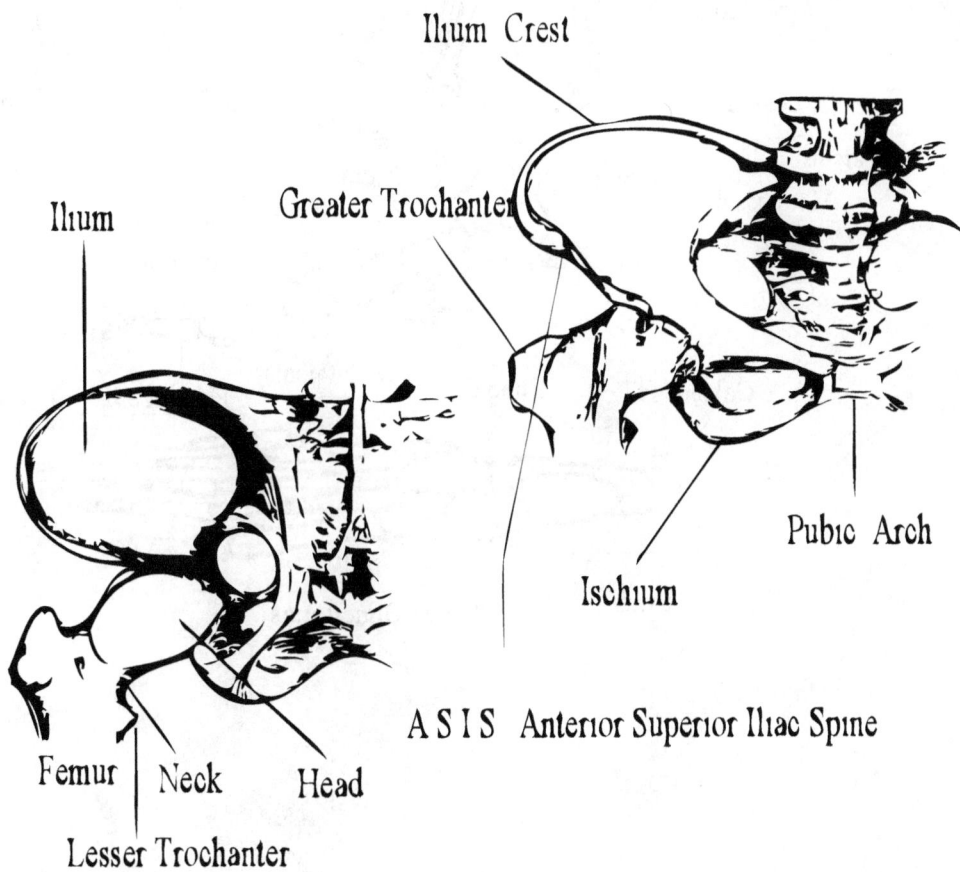

Ilium Crest

Ilium

Greater Trochanter

Pubic Arch

Ischium

Femur   Neck   Head

Lesser Trochanter

A S I S Anterior Superior Iliac Spine

60 BONES FORM THE LOWER LIMBS (30 in each leg):

- Femur
- Patella
- Tibia
- Fibula
- Tarsal bones x 7
- Metatarsal bones x 5
- Phalanges x 14

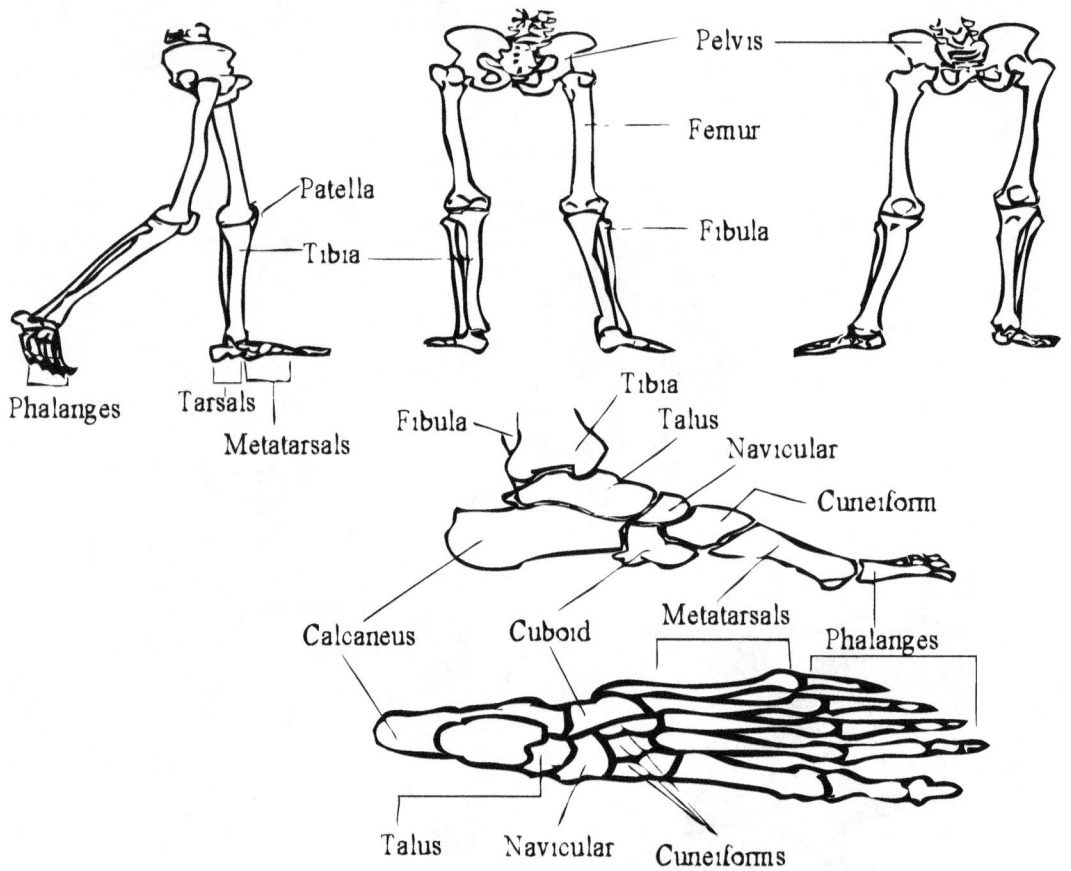

Problems Related to the Skeletal System

There are four main causes of spinal curvature:

1. Congenital: This is an exaggerated curvature of the spine which is present at the time of birth.

2. Traumatic: This is an exaggerated curvature of the spine resulting from an accident.

3. Environmental: This is an exaggerated curvature of the spine resulting from bad posture.

4. Hereditary: Due to a genetic weakness from one generation to another.

Exaggerated Curvature of the Spine Falls into Three Categories:

1. Kyphosis: Exaggerated outward curvature of the spine in the thoracic region.

2. Lordosis: Exaggerated inward curvature in the lumbar region.

3. Scoliosis: Exaggerated lateral curvature of the spine in any area.

A classic example would be "The Hunchback of Notre Dame" who had kyphosis and scoliosis of the thoracic region.

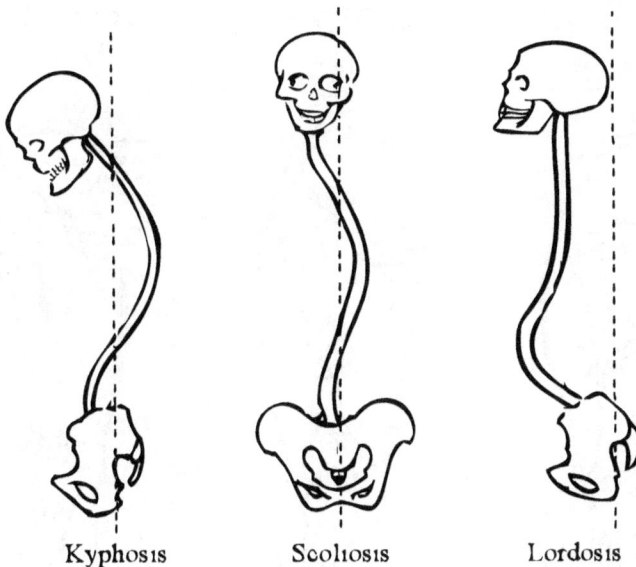

Kyphosis   Scoliosis   Lordosis

There Are Six Main Classifications of Bone Fractures:

1. Compound fracture: When the bone is broken and protrudes through the skin.
2. Simple fracture: A single break with no serious damage to surrounding tissue.
3. Greenstick fracture: This is an incomplete fracture. (as seen in children).
4. Complicated fracture: When the broken bone causes damage to surrounding tissue.
5. Comminuted fracture: When the bone breaks in a number of places.
6. Impacted fracture: When the broken end is driven into the other broken end.

Compound   Simple   Greenstick

Complicated Comminuted Impacted

Some Conditions of the Skeletal System

Disease or Disorder **Arthritis**
Description An inflammation of a joint.

There are several different types of arthritis, but, **osteoarthritis** and **rheumatoid arthritis** are the two you will come across most often.

Osteoarthritis; is a form of arthritis occurring in one or more joints which are subject to degenerative changes, causing loss of articular cartilage, and increasing bone and cartilage mass in the joint, medically this is termed "osteophytes". Inflammation of the synovial membrane of the joint is common late in the disease. There are many causes that will initiate arthritis ie. Congenital or genetic defects, infection, metabolic or endocrine diseases and trauma to name but a few.

Rheumatoid arthritis; A chronic, destructive and often deforming disease, characterized by inflammation of the synovial fluid leading to swelling of the joint. More commonly found in females and is variable in nature by frequent periods of remissions followed by as many periods of exacerbations. Considered by many doctors to be caused by auto-immune complications. In chronically affected joints, the delicate synovial membrane develops many villous folds and thickens because of increased numbers and size of the synovial lining cells and colonization by lymphocytes and plasma cells.

Possible Allopathic Rest, heat pads, steroidal and non-steroidal anti-inflamma-
Treatment tory drugs (NSAIDs) and pain killing drugs. Intra-articular corticosteroids are often used in severe cases and have the most dramatic effect short term. However, corticosteroids remain active within the body for years. Surgical intervention is often necessary. Range of motion exercises and good diet often cause improvement.

Disease or Disorder Description	**Osteomyelitis**
	Infection of the bone leading to abscess formation. Inflammation of the bone marrow, usually caused by bacteria or fungi, introduced by trauma (open fractures) or contamination during surgery. The bacteria called Staphylococci is the most common causative agent. The long bones in children and the spine in adults are the most common sites of infection. Persistent and severe pain and tenderness are associated with this disorder. In patients with diabetic or atherosclerotic arterial insufficiency of the lower limbs, organisms reach the bone by entering the soft tissue through a cutaneous foot ulcer. Osteomyelitis of the skull typically arises from sinus or dental infections.
Possible Allopathic Treatment	Bed rest and parenteral antibiotics or appropriate antimicrobial agents are administered, and as a last resort, surgery may be necessary to remove necrotic bone. The main indication is drainage of sub-periosteal or subcutaneous abscesses.
Disease or Disorder Description	**Osteoporosis**
	Generalized loss of bone tissue, calcium salts and collagen in bone. This occurs during ageing, and most frequently in post-menopausal women. Osteoporosis may be idiopathic or secondary to other disorders, such as thyrotoxicosis or the bone demineralization caused by hyper-para-thyroidism. It is characterized by excessive loss of both calcified matrix and collagenous fibres from the bone, resulting in a dangerous pathological condition of bone degeneration, increased susceptibility to deformity of the spine and/or "spontaneous fractures".
	Please note: Great care must be extended when dealing with this disorder: do not exert pressure on any joint. The School does not suggest that you attempt to treat this disorder with manual therapy.
Possible Allopathic Treatment	Treatment should involve preventative measures by assessing the risk factors, ie. Postmenopausal women may be given estrogen replacement therapy. Estrogen may arrest or decrease disease progression. Calcium supplements may show increased bone density under certain dietary conditions.

Disease or Disorder	**Osteosclerosis**
Description	An abnormal increase in the density of bone tissue with thicker, harder bone. This is sometimes due to a congenital disorder. But is commonly associated with ischemia (a decreased oxygenated supply of blood to an organ), chronic infection or tumour formation, and may also be caused by faulty bone reabsorption as a result of some abnormality involving the osteoclasts (growth and repair during bone healing).
Possible Allopathic Treatment	Treatment is symptomatic, requiring rest and relief from pain.

Disease or Disorder	**Ankylosing spondylitis**
Description	There are two types; 1. spondylitis and 2. ankylosing spondylitis Ankylosing spondylitis is the most severe form of arthritis of the spine. Similar to rheumatoid arthritis in symptoms, but different in that males are affected more than females. The articulating cartilage is destroyed, fibrous adhesions develop, and eventually bone fusion occurs with calcification of the intra vertebral discs. This condition often occurs in the sacroiliac joints and spreads slowly upward.
Possible Allopathic Treatment	Joint discomfort is relieved by pain killing drugs, non-steroidal anti-inflammatory drugs (NSAIDs), exercise and other supportive measures by suppressing articular inflammation, pain, and spasm. As a last resort, surgery may be required.

Terminology

The Anatomical Position	This is the human body in the erect position with arms by the sides and the palms facing forward.
Adduction	Brings a limb medially towards the midline of the body
Anterior	This applies to the front of the body while in the erect position.
Appendicular Skeleton	The upper and lower limbs and their girdles.
Axial Skeleton	The skeleton of the head and trunk.
Bursa	Small sac filled with fluid surrounding a joint.
Cancellous Tissue	Characterized by latticed structure as seen in the spongy tissue in bones.
Cartilage	A substance similar to bone, but not as hard. It acts as a cushion between bones and it gives shape to the nose and ears.
Dislocation	Displacement of a joint. It occurs when force is applied to the joint capsule causing the bones to separate, and this applies particularly to the ball and socket joints, as the ball is forced out of the socket. When dislocated bones are returned to their proper position, it is known as "reduction".
Distal	A point some distance from a given named location.
Dorsal	Anything that is related to the back and is the term which is normally used when describing the back, hand or foot, eg. (back of the hand or upper part of the foot).
Dorsiflexion	To bend the ankle/foot towards the shin (like lifting-off the gas pedal).
Eversion	Moves the sole of the feet laterally.
Extension	To straighten a limb or open a joint, should it extend beyond its range of motion is called "hyperextension".
Flexion	A movement that bends a joint or that brings the bones closer together.
Foremen	Hole in a bone.
Gouty Arthritis (Gout)	Can occur in any part of the body, but is popularly associated with the big toe, resulting from urate crystals (chalky salts of uric acid) being deposited in and around the cartilage. This form of arthritis is much more common in men than in women.
Inferior	Situated below a structure.

Inversion	Brings the sole of the foot medially (in).
Inferior	Situated below.
Lateral	This is to either side of the median line, eg. the outer side of the arm will be the lateral aspect while the inner side will be the medial aspect.
Medial	This is an imaginary line which runs straight through the centre of the body, from the top of the head down between the two feet. Refers to structure closest to the midline of the body.
Orthopaedics	A branch of surgery concerned with corrective treatment of the skeletal system and the study of bone disease.
Osteo	Referring to bone.
Osteology	The study of bones.
Osteoblasts: (Os-te-o-blasts)	When the skeleton first forms in a fetus, it consists not of bones but of cartilage and fibrous structures shaped like bones. Gradually, these cartilage "models" become transformed into real bones when the cartilage is replaced with calcified bone matrix deposited by specialized bone-forming cells called "Osteoblasts".
Palmar	Anything that concerns the palm of the hand.
Periosteum	A hard membrane adhering to bone and providing a protective cover. It contains blood vessels supplying blood to the bone and in its deepest layers are the bone-forming cells called "Osteoblasts".
Plantar	The underside of the foot (the sole). Planta flexion, to point your foot towards the floor like stepping on the gas pedal.
Posterior	This is the back of the body when in the erect position.
Proximal	This is a term of comparison applied to a structure which is nearer the centre of the body or median line, eg. proximal thigh is the end of the thigh nearest to the centre of the body.
Pronation	Normally used to describe the pivoting action of the upper limbs to face the palm down.
Rheumatology	Study chronic muscle and joint diseases.
Sesamoid	Small nodule of bone that often forms in the tendons at pressure points; a small rounded mass of bone, ie. Patella.
Skeleton	Articulated bony framework of the body.

Spondylitis	A type of arthritis which attacks the spinal vertebrae. The severest form of this disease is called "Ankylosing Spondylitis", where bone and cartilage fuse together, resulting in complete immobility.
Supination	Normally used to describe the pivoting action of the upper limb to face the palm upwards.
Superior	Above or on top of.
Superficial	This term describes a structure closest to the surface.
Thorax:	Chest compartment enclosed by the ribs, backbone and diaphragm.

The Muscular System

Definition: Kinesiology - " the scientific study of movements"

When referring to the skeletal system or the muscular system we must also study the movements of muscles as well as the anatomical structure and position of this system.

Historical

The history of this system goes back many years. Philosophers and scientists like Hippocrates, Aristotle and Galen and other gifted men supplied many important texts on this subject. Artists like DaVinci, Galvani and Vesalius have depicted and illustrated many incredibly accurate drawings that have captured the imagination causing such a fascination that even to this day, modern anatomy teachers frequently use their work to emphasize a point.

What is a Muscle

Muscles are composed of highly specialized cells that can generate a focused force or produce skeletal movement. Skeletal muscle cells are responsible for all voluntary actions that range from speaking to moving the limbs. Cardiac and smooth muscles are involuntary muscles that maintain the functions of the cardiovascular, respiratory, digestive, gastrointestinal, and genito-urinary systems. Muscles are responsible for up to 50% of the total body weight. **Their function is to permit movement of the skeleton.**

There are two types of muscle:
- **Voluntary muscles which are striated (cross banded)**
- **Involuntary muscles which are smooth**

Voluntary muscle

These are muscles that are under conscious control. Muscles attached to the skeleton are called voluntary muscles. This means that they are anatomically and mechanically arranged in parallel, to allow the cells to function independently of one another. The total force exerted will equal the sum of the force generated by each cell. Voluntary muscles are striated in appearance and a typical striated muscle is composed of thousands of muscle fibres enclosed in bundles surrounded by a sheath of fibrous tissue called a *"fascia"*. At either end of this muscle is a whitish coloured, non-contractile fibrous tendon which serves to attach the muscle to the bone, cartilage, ligament or other tendinous structure.

There are four types of structure of striated muscle:

1. **Fusiform**: The fibre bundles lie almost parallel to the long axis or "line of push or pull" of the muscle, but they are slightly curved, so that the muscle tapers at each end.
2. **Unipenniform**: The muscle bundles are parallel to each other but the fibres lie obliquely and converge, like the plumes of a medieval quill pen, to one side of a tendon.
3. **Bipenniform**: Similar to Unipenniform but with a larger number of fibres that runs down both sides of the muscle with a tendon that runs the complete length of the muscle down the centre, similar in appearance to a feather. An example would be the Rectus femoris.
4. **Multipenniform**: Similar to Bipenniform but with an even larger number of fibres. These fibres are arranged in the curved bundles in one or more planes as found in Sphincter muscles.

Involuntary muscle

These are muscles that operate without conscious control, and are responsible for the life-preserving functions of the body. e.g. heart, respiration, digestion, etc. These muscles must be capable of not only generating force and movement, like skeletal muscle cells, but of also maintaining organ dimensions against applied loads, ie. vascular smooth muscles must bear the load imposed by the blood pressure to regulate blood flow.

Cardiac Muscle

There is however, a third category or type of muscle which is a mixture of the two. This is known as "cardiac muscle". A cross section of the cardiac muscle shows that while it is an involuntary muscle, it also displays the characteristics which bear a superficial resemblance to a voluntary muscle, although the striation is not as well defined.

Points of Attachments:

The term *"the point of origin"* is meant to convey its fixed or central point of attachment. The term *"the point of insertion"* is meant to convey the moveable point to which the force of the muscle is directed. As a rule, the *origin* is the end of the muscle proximal to the centre of the body and the *insertion*, distal to the body.

Muscle consists of a number of **contractile or elastic fibres** bound together in bundles. These are contained in a thick band which is usually spindle shaped and always contained in a sheath. The sheath extends to form strong fibrous bands known as tendons, and it is the tendons that attach the muscle to the bone. These bundles are termed *"myofibrils"*.

Muscles receive a stimulus, **noradrenaline**, one of the chemicals produced by the autonomic nervous system in nerve transmission to the muscle from motor nerves. In response to this stimulus, the muscle is shortened.

Muscles always work in pairs. The one that contracts is known as the *"synergist"* and is the prime mover , while the muscle that extends is known as the *"antagonist"*.

Example of an EXTENSOR MUSCLE

TRICEPS An extensor extends a limb

This action and reaction between the synergist and the antagonist means that the muscles are never completely at rest, being either slightly under tension or slightly under contraction. This tension and contraction is called muscle tone. When we bend our forearm the biceps contract (synergist), while the triceps extend (antagonist). However, to straighten the arm, the biceps then become the antagonist and the triceps become the synergist.

Example of a FLEXOR MUSCLE

BICEPS   A flexor flexes a limb

How muscles move:
Skeletal muscle movement is initiated by a different chemical, "acetylcholine". One of the major achievements of the 20th-century is the discovery of what makes a muscle work. The process involves a chemical reaction and an exchange of nerve signals. For instance, when your arm dangles by your side, the biceps muscle appears thin, and stringy. But if you clench your fist, and flex your forearm, the biceps muscle becomes tense and bulges out.

To help you understand this principle, imagine a small room with movable walls, representing a sarcomere, or unit of muscle contraction. Now, imagine you are standing in the middle of that room. You are now in the position of a myosin filament and in front of you are two ropes. One attached to the left wall and the other attached to the right wall. These represent thin acting filaments. In muscle contraction, you would be stimulated to pick up the two ropes and pull on them, bringing the walls closer. In relaxation, you'd drop the ropes, and the walls would slide back. What keeps them locked together?   The main factors are nerve signals that set off chemical reactions that contribute to the process of contraction.

The point where the message-bearing nerve-fibre links with the muscle fibre is called the "motor end plate". This activating mechanism lies within the muscle fibre. When a message reaches the end plate, the plate secretes the powerful chemical called acetylcholine, which passes into the muscle fibre and produces jolting electrical charges that get the muscle action under way.

There are some 640 named muscles in the body, but there are thousands of muscles that are unnamed. When a muscle contracts, energy is required. This involves a breaking down of glucose, glycogen and fat, which in turn liberate the energy required for movement. Whenever energy is used during this process, there will always be waste products excreted from the muscles to be taken away by the venous system.

Should the activity produce more waste products than the venous or lymph systems are able to cope with, (these waste products being "lactic acid, carbon dioxide, heat and water"), then the waste remains in the muscle fibres and gives a feeling of stiffness. During this increased activity, the heart rate quickens to increase the flow of blood to the muscles to take away these metabolites and to help dissipate the heat generated.

Muscles are put into groups depending on the function they perform.

- An adductor bends a limb towards the median line.
- An abductor takes a limb away from the median line.
- A sphincter surrounds and closes an orifice or opening.
- A supinator turns a limb to face upwards.
- A pronator turns a limb to face downwards.
- A rotator rotates a limb.

Terminology

atony	loss of muscle tone
atrophy	muscle wasting or a reduction of muscle size
ganglion	a cystic swelling of a joint or tendon sheath. Commonly found on the back of the wrist
myositis	a term used to indicate muscle inflammation
sprain	an injury to a ligament
strain	an injury to a muscle or tendon

The following is a short list of some of the principal muscles of the body. **This is not a comprehensive list of all muscles**, only those most commonly used and named. However, the student is encouraged to study this subject in greater depth at a later date.

**Muscles of the body
Anterior view**

**Muscles of the body
Posterior view**

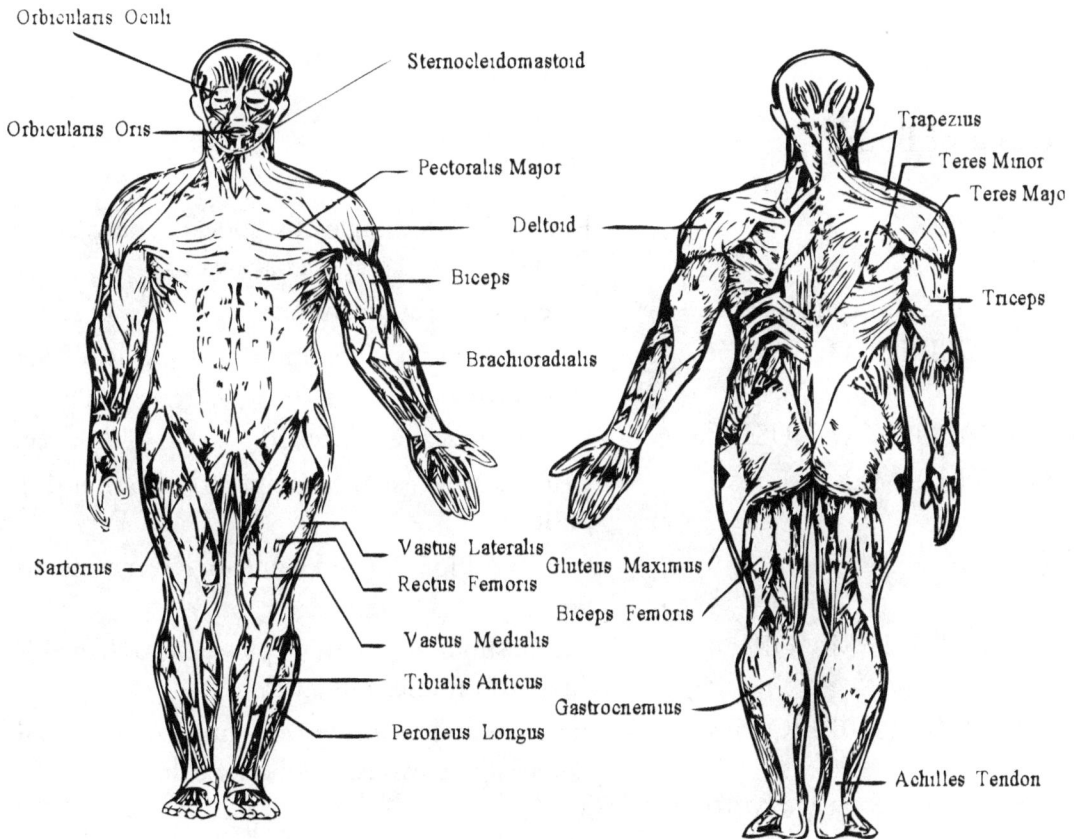

Orbicularis Oculi

Orbicularis Oris

Sternocleidomastoid

Pectoralis Major

Deltoid

Biceps

Brachioradialis

Sartorius

Vastus Lateralis

Rectus Femoris

Vastus Medialis

Tibialis Anticus

Peroneus Longus

Trapezius

Teres Minor

Teres Majo

Triceps

Gluteus Maximus

Biceps Femoris

Gastrocnemius

Achilles Tendon

Muscles of the Body

Head and Neck

FRONTALIS (or epicraneas)	Elevates eyebrows and draws scalp
ORBICULARIS OCCULI	Closes eyelids.
ORBICULARIS ORIS	Puckers mouth.
MASSETER	Muscle of mastication.
BUCCINATOR	Compresses cheeks and retracts the angle of the mouth.
STERNO-CLEIDO MASTOID	Flexes head and turns head from side to side.
PLATYSMA	Muscle of facial expressions.

Trunk of Body

TRAPEZIUS	Rotates inferior angle of scapula laterally, raises shoulder, draws scapula backwards.
SPINALIS (or spinatus)	Extends vertebral column.
LATTISIMUS DORSI	Adducts the shoulder and draws the arm back and down.
SERRATUS MAGNUS	Draws the scapula forward.
GLUTEUS MAXIMUS	Extends to assist in raising body from sitting position.
PSOAS	Flexes hip joint and trunk on lower extremities.
PECTORALIS MAJOR	Flexes shoulder joint, depresses shoulder adducts and rotates humerus.
ABDOMINIS OBLIQUES	Supports abdominal viscera and flexes vertebral (internal and external oblique) column.
ABDOMINIS TRANSVERSALIS	As above
ABDOMINIS RECTUS	As above

Arms and Legs

DELTOID	Abduction of the humerus to right angles.
BICEPS BRACHIALIS	Flexes and supinates forearm.
TRICEPS BRACHIALIS	Extends elbow joint.
BRACHIALIS ANTICUS	Flexes elbow joint.
CORACO BRACHIALIS	Flexes and adducts humerus.
BRACHIO RADIALIS (supinator longus)	Flexes elbow joint.
PRONATOR TERRES (Pronator radii terres)	Pronates forearm.
SUPINATOR (Supinator radii breves)	Supinates forearm.
FLEXOR CARPI RADIALIS	Flexes wrist joint.
RECTUS FEMORIS	Extends knee joint
VASTUS LATERALIS	Extends knee joint
VASTUS MEDIALIS (Quadriceps)	Extends the knee joint.
SARTORIUS	Flexes hip and knee joints and rotates femur.
ADDUCTOR MAGNUS	Adduct thigh.
LONGUS	Adduct thigh
AND BREVIS	Adduct thigh
BICEPS FEMORIS (Ham strings)	Flexes knee joint.
SEMI-TENDINOSUS (Ham strings)	Flexes knee joint and extends hip joint.
SEMI-MEMBRANOSUS (Ham strings)	Flexes knee joint and extends hip joint.
GRACILLIS	Adducts femur and flexes knee joint.
GASTROCNEMIUS	Flexes ankle and knee joint.
TIBIALIS ANTICUS	Flexes and inverts foot.
PERONEUS LONGUS	Inverts and flexes foot and supports arches
FLEXOR DIGITORUM LONGUS	Flexes toes
EXTENSOR DIGITORUM LONGUS	Extends toes
TENDON ACHILLES	Assists in flexion of the foot.

Some Conditions of the Muscular System

Disease or Disorder   **FIBROSITIS**

Description   Inflammation of the fibrous connective tissue, usually characterized by a poorly defined set of symptoms, created by a build up of urea and lactic acid. It usually, normally includes, pain and stiffness of the neck, shoulder and upper body. Fibrositis usually developed in middle age. The person is often tense, leading to the belief of a psychosomatic or psychogenic origin (stress).

Possible Allopathic Treatment   Salicylates, sedatives, tranquillizers, muscle relaxants and intra-articular injections of a local anaesthetic may be prescribed intreatments

Disease or Disorder   **LUMBAGO (low back pain)**

Description   Fibrositis of the muscle in the Lumbar region, which is often referred to today as fibro-myalgia. This is where a group of common non- articular rheumatic disorders, characterized by achy pain, tenderness and stiffness in the muscles or tendons at the point of insertion. Often related to overuse or microtrauma causative factors. More common in females than males, and may be caused or intensified by stress. Men on the other hand often show or develop localized pain due to occupational or recreational strain.

Possible Allopathic Treatment   In milder cases this disorder may remit spontaneously with the reduction of stress. Stretching exercises may help, better sleep patterns, local application of heat and gentle massage. Low doses of tricyclic agents at night (amitriptyline) will improve sleep, and aspirin to help manage pain. Painful areas may respond to injections of a 1% Lidocaine solution or hydrocortisone acetate. Even antidepressant drugs at a very low dose may prove useful.

Disease or Disorder	**TORTICOLLIS**
Description	Fibrositis of the neck muscles (Sterno-Cleido Mastoid muscle). The head takes up an abnormal position due to the neck muscles being in a state of contraction. Believed to be associated with injury to the sternocleidomastoid muscle on one side at birth. In adult life it is normally the result of trauma or a secondary disorder associated with disorders like Tardive Dyskinesia, hyperthyroidism, basal ganglia disease etc.
Possible Allopathic Treatment	Normally treatable but neurological and idiopathic processes are more difficult to treat. The spasm responds well to physiotherapy. Drugs are effective for pain (ie. anticholinergics) muscle relaxants (ie. caclofen) or tricyclic antidepressants (ie. amitriptyline). Multiple injections of botulinum toxin type A (oculinum) into the dystonic muscles of the neck may improve the position of the head or reduce painful muscle spasm. On a surgical approach, selective denervation of the neck muscles shows the most success.
Disease or Disorder	**CRAMP**
Description	A localized painful area effecting one or more of the muscle groups. The most common cause being either vigorous exercise or certain metabolic disorders. e.g. when there is a sodium or water depletion. Severe cramps normally occur in striated muscle, resulting from excessive exertion and/or high ambient temperatures causing sweating. Although cramps are often the result of excessive loss of sodium and occasionally potassium and magnesium during strenuous activity at high atmospheric temperatures (38C), the onset is often abrupt, affecting the muscles of the extremities first. Severe pain and carpo-pedal spasm (a spasm of the hand or foot) may temporally incapacitate the muscles of the hands or feet. Often sporadic, the cramping makes the muscles feel like hard knots. Vital signs are usually normal. The skin may be either hot and dry or clammy and cool.
Possible Allopathic Treatment	In most instances, cramps are prevented or are rapidly relieved by drinking fluids or eating foods containing sodium chloride (salt). If the patient cannot take food or drink orally, 0.9% sodium chloride IV may be necessary in extreme circumstances. Sodium chloride tablets are often used as a preventative measure, but may cause stomach irritation, and an overdose may lead to oedema. Awareness of the problem is usually sufficient to prevent it.

Disease or Disorder	**POLIOMYELITIS**
Description	Commonly called "Polio" which arises in the neurological system along with multiple or disseminated sclerosis which also arises in the neurological system, though both of these disease conditions can also profoundly effect the bodies musculature. This disease is infectious and is transmitted by faecal contamination or oropharyngeal secretions. The severity of the infection increases with age.
Possible Allopathic Treatment	Treatment of the abortive and non-paralytic forms of this disease is by treating the symptoms, plenty of bed rest, and good food to lower stress, and fatigue. Treatment of the paralytic form will include a stay in hospital under observation. The application of hot packs and baths together with physiotherapy will improve the range-of-motion.

Disease or Disorder	**FIBROMYALGIA**
Description	By the time this disorder is diagnosed, it is often chronic, causing inflammation of muscle tissue, more commonly found in voluntary muscle. Characterized by aches and pains, tenderness and/or stiffness, it is often localised around areas of tendon insertions. Fibro, means fibrous tissues that include muscles, tendons, ligaments, and other "white" connective tissues and myalgia means, muscular pain. This condition is more common in females and is made worse by damp or cold conditions, physical or mental stress and sleeping disturbances. Men are prone to developing localized fibromyalgia associated with occupational or recreational strain. Symptoms can be exacerbated by physical or emotional stress.
Possible Allopathic Treatment	Fibromyalgia has been known to remit spontaneously in mild cases, but will become chronic if stress is not dealt with. Stretching exercises will help, sleeping pills are often prescribed, heat pads will offer some relief as will gentle massage. Low doses of tricyclic drugs like amitriptyline will help with sleep, Aspirin may help with pain. Incapacitating local areas of pain may be injected with 1% lidocaine solution or hydrocortisone acetate suspension.

Disease or Disorder	**BURSITIS**
Description	Acute/chronic inflammation of the bursae, also involving the connective tissue and surrounding structure of a joint. Severe pain is the chief symptom, particularly during movement of the afflicted limb.
Possible Allopathic Treatment	Pain control with drugs and maintenance of joint motion is the main concern. A popular measure used for pain relief is intrabursal injections of adrenocorticosteroid, analgesics, anti-inflammatory agents, ice or cold pads and immobilization of the inflamed joint.

Disease or Disorder	**MUSCULAR DYSTROPHY**
Description	Muscular dystrophy (MD) refers to a genetically transmitted disease characterized by a progressive atrophy of skeletal muscle groups without evidence of degeneration of neural tissue (unlike MS). Frequently found in young boys aged 3-7 yrs, the disease is characterised by loss of strength, disability and even deformity. The cause is still unknown but evidence leads us to believe that it is linked to an inborn error of metabolism. The pelvic girdle is effected first and then the shoulder girdle. There are two main types of MD, "pseudohypertrophic" (duchenne) limb-girdle muscular dystrophy and "facioscapulohumeral" (landouzy-dejerine) characterized by weakness of the facial muscles.
Possible Allopathic Treatment	Diagnostic confirmation is made by muscle biopsy, electromyography and genetic pedigree. Treatment consists of physiotherapy and orthopaedic procedures to minimize deformity. The added burden of obesity should be avoided. Prednisone may improve functional capabilities.

Disease or Disorder	**ANKLE SPRAINS**
Description	An injury to a ligament. Ankle sprains can be classified according to the extent of soft tissue damage. grade 1: Mild, no ligament tearing, tender area with minimum swelling. grade 2: Moderate, partial or incomplete tearing, painful and walking with difficulty, obvious swelling. grade 3: Severe, complete tear of the ligament, swelling, internal haemorrhaging, walking is out of the question, very painful.
Possible Allopathic Treatment	grade 1: strapping with elastic bandages, ice and elevation of the limb. grade 2: a walking-cast to immobilize the ankle for three weeks. grade 3: cast immobilization or surgery. Surgery is controversial because the extreme fragmentation of ligaments makes repair difficult. Some surgeons cast solitary anterior talofibular ruptures but recommend surgical repair if the fibulocalcaneal ligament is torn. Ice-packs to reduce swelling and anti-inflammation.

Disease or Disorder	**STRAIN**
Description	An injury to a muscle or its tendon. Caused by exerting physical force resulting in injury to the muscle or tendon.
Possible Allopathic Treatment	Strap with elastic bandages, ice if bruising is present and elevate the limb. Healing is rapid 2-4, days if rest is observed. Ice-packs to reduce swelling and anti-inflammation.

The Digestive System

Introduction

The digestive system could be described as an assembly line that works in reverse, because it takes in whole food products and breaks them down into their chemical components. The food you eat is broken down by digestive juices into small, easily absorbable nutrients that generate the energy required to maintain a healthy body.

For a healthy and efficient working body to be maintained, the food content needs to be both balanced and nutritious. We are to a greater or lesser degree responsible for our own basic health; the old saying "we are what we eat" is true in many cases. Because of this, we need to have at least a basic understanding of the process involved with digestion and to understand some of the ways in which it can go wrong.

The body requires raw material to grow and repair itself as well as for heat and energy. These raw materials are supplied in the form of food that comes in a variety of packages that we ingest and convert to compounds which generate and sustain life.

It takes about three to six hours for a meal to be converted from solids to semi-liquids in the stomach. The rate at which the stomach moves food along is controlled primarily by the duodenum. The sphincter releases hormones that control the muscle movements of the stomach regulates the rate of digestion. As a result, the duodenum receives the chyme gradually, in just the right amount for optimum digestion and absorption. When the stomach is full, it signals for the release of the hormone gastrin, which speeds up digestion.

The average adult has a digestive tract that is about thirty feet long from one end to the other, it could be described as a strong muscular continuous tube, starting at the mouth and passing through the larynx, oesophagus, stomach, small and large intestine and ending at the rectum.

Associated with this system are the accessory organs:

- the tongue
- teeth
- salivary glands.
- liver
- pancreas

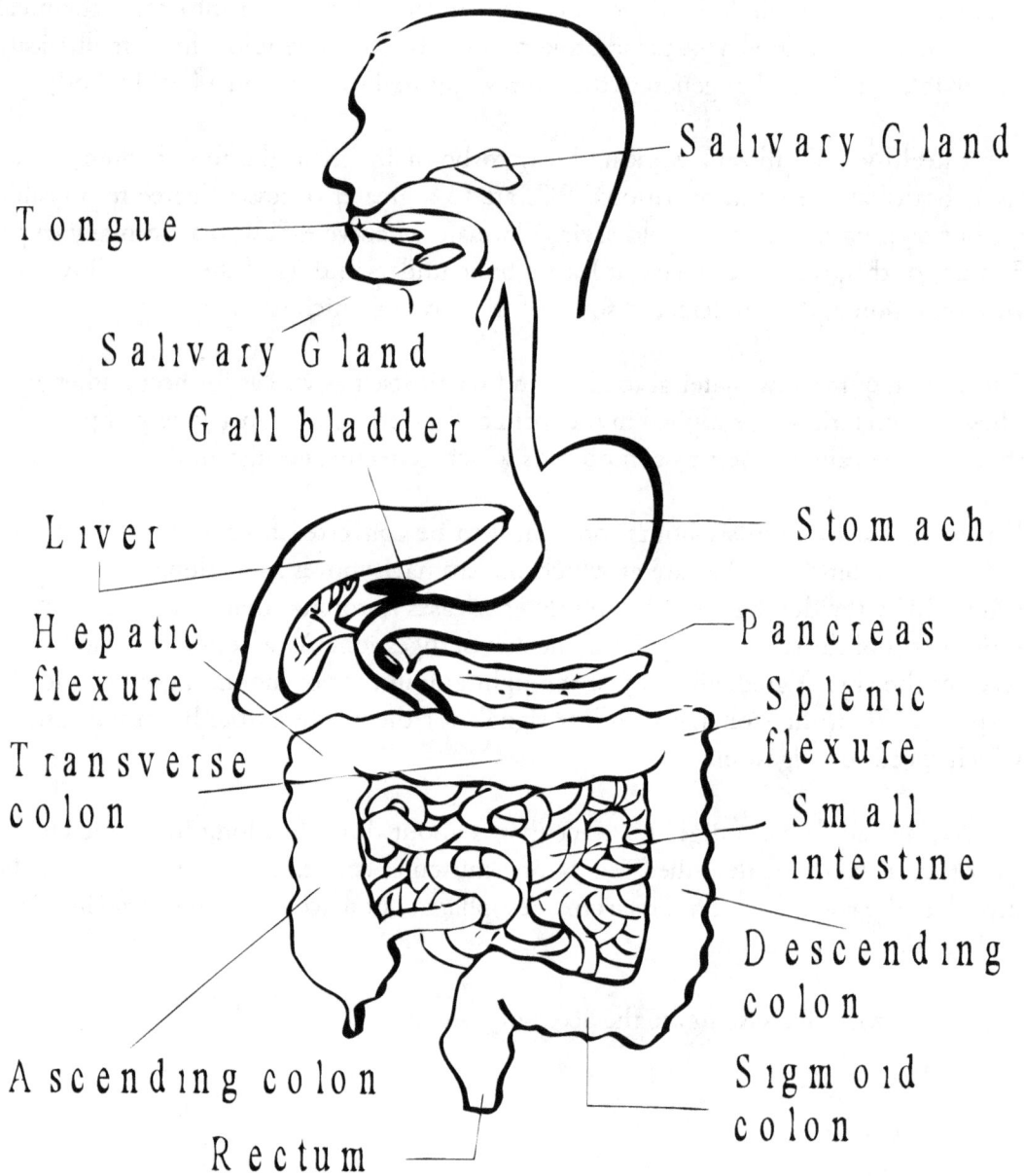

Salivary Gland

Tongue

Salivary Gland

Gall bladder

Liver

Stomach

Hepatic
flexure

Pancreas

Splenic
flexure

Transverse
colon

Small
intestine

Ascending colon

Descending
colon

Rectum

Sigmoid
colon

Lets start at the beginning.

The Mouth
It is the body's first line of defence for the digestive system.

The mouth has at least four functions.
1. To break up food by chewing
2. To lubricate with saliva
3. To assist in regulating the body temperature
4. To consciously initiate swallowing

The Tongue:
The tongue is an immensely mobile mass of muscle which helps the teeth to tear food to pieces by forcing it against the bony top plate of the roof of the mouth, and then form the crushed and moistened particles into a ball, or bolus ready for swallowing. When the tongue moves up and back, pressing against the hard palate, it propels the bolus to the back of the mouth and into the oesophagus. It also gives warning of possible injury by registering pain when foods are too hot, and revulsion when they are spoiled. It consists of striated voluntary muscle and it is attached mainly to the mandible and hyoid bones.

The upper surface of the tongue is covered with papillae, of which there are three forms:

- The filiform papillae, found chiefly on the dorsal aspect of the tongue.
- The fungiform, found mainly on the sides and tip of the tongue.
- The vallate, which is the largest of the papillae in a "v" formation at the back of the tongue. Taste-buds are found in the walls of the vallate papillae.

On a healthy tongue, the papillae are usually pinkish-white and velvety smooth, and are crossed by slits or fissures that reveal the red tongue beneath. Although they combine to make a myriad of different flavours, there are believed to be only four primary tastes; sweet, sour, bitter and salty.

Various parts of the tongue have taste buds that are especially sensitive to one of the basic sensations of taste. You taste salty and sweet mainly on the front of the tongue, sour on the sides and bitter at the back. The middle of the tongue registers almost no taste. As a matter of interest, there is a substance that will register sweet on the tip of the tongue and bitter by the time it reaches the back, this is "saccharin".

Taste is highly individual. This means that different people will be attracted to different foods and flavour, part of the explanation being heredity. Some genes make the receptors for bitterness especially sensitive. In these people, saccharin is likely to produce a strong sensation of bitterness. Each person's saliva has its own special taste, which in turn effects the taste of the food. If, for example, your saliva has a low sodium content, food containing a given amount of salt would taste saltier to you than to another person whose saliva is high in sodium.

The Teeth:

There are thirty-two permanent teeth in the adult human body.

Deciduous teeth start to appear at about six months, by the age of three the child has about twenty teeth. Between the ages of six and twelve the deciduous teeth are shed and replaced by permanent teeth. A further six teeth and molars will appear. By the age of 25 a total of 32 teeth should be present. 8 Incisors, 4 canine, 8 premolars (or bicuspids) and 12 molars.

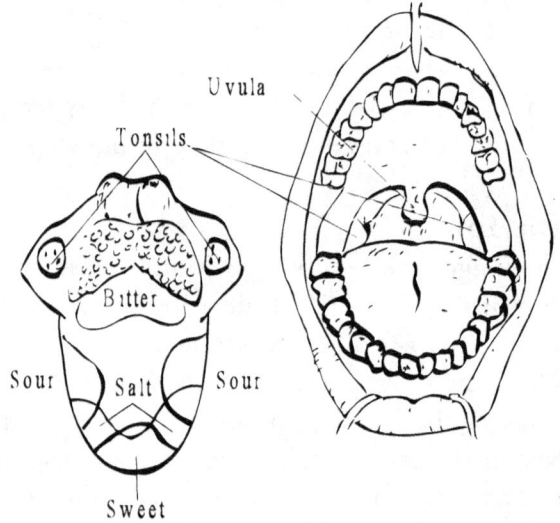

The tooth consists of three parts:

The crown:

This consists of a dense mineral or enamel surrounding the hard dentine, which has a soft centre called "the pulp". This is filled with blood vessels, lymphatics and the nerves, which reach it through the root canal.

Neck:

The neck adheres to the gum.

The root:

The root penetrates the bone, where it is held in place by a ligament and cementum.

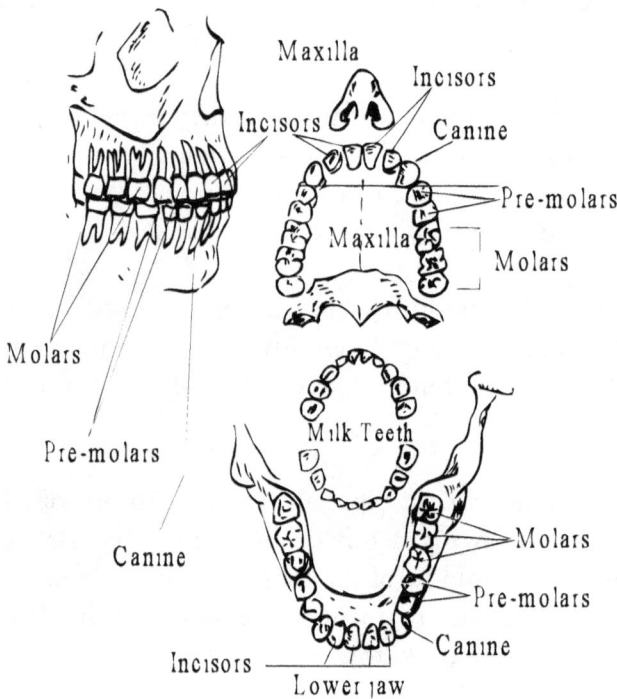

Salivary Glands:

There are three pairs of salivary glands:

1. The parotid glands, found in front of and a little below the ears. You may have become familiar with this gland when you were a child as these were the glands that swelled when you had mumps, swollen or not these are the largest salivary glands.
2. The sublingual glands, situated just below the tongue, are smaller than the parotid glands, about the size of a walnut.
3. The submandibular glands, situated below the mandible, are the smallest of the salivary glands.

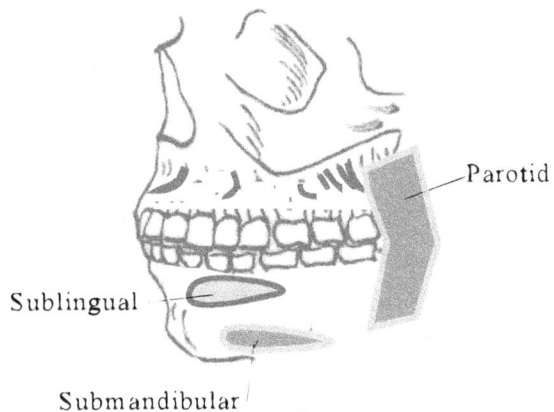

The salivary glands produce secretions containing the starch-reducing enzyme, ptyalin which helps in the digestion of cooked starches.

To help you get a clear picture of what saliva does, try chewing on a piece of dry bread, paying attention to the way it tastes. When it develops a sweet flavour, you will know that your saliva has begun to convert the complex starch molecules in the bread to a mixture of simple sugars, glucose and maltose. This is accomplished by a digestive enzyme known as salivary amylase, or ptyalin. Of course, saliva does a lot more than just break down starches into sugars, because without saliva you would find it very difficult to swallow. The mucus in saliva adheres to food and moistens it during chewing to enable it to slide easily down the oesophagus. Another function of saliva is to help keep the mouth healthy because it is also mild germicide, saliva kills bacteria, especially the kind that causes so much tooth decay.

From the mouth the food then passes into the pharynx which is a muscular tube with seven openings. These are:

1. The mouth
2. The oesophagus
3. The larynx
4 & 5. Two posterior apertures of the nose
6 & 7. Two auditory (eustachian) tubes from the ear

From the pharynx the food passes into the oesophagus, which is a muscular tube lined with mucous membrane and covered with fibrous tissue. From here, the food passes into the stomach.

The Stomach

The stomach is a muscular sac. Its size and shape varying with its contents and muscular tone. When it has no food in it, it is shaped like a "j"; when it is full, it takes the shape of a boxing glove. When filled to capacity, the average stomach can hold up to 1.5 litres of food.

The stomach presents two curvatures, the greater and lesser curvature and for the purpose of description, is divided into three parts;

1. The cardiac portion
2. The body
3. The pyloric

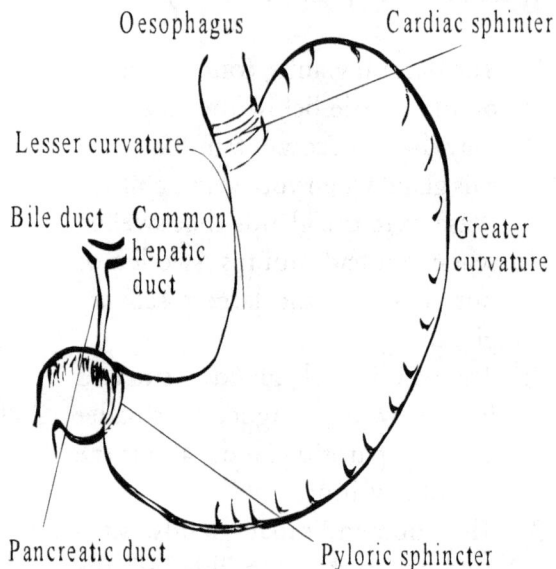

The openings into the stomach are guarded by circular bands of muscle, like purse strings. The cardiac sphincter muscles at the top end, and the pylorus sphincter muscle at the other end. The cardiac sphincter prevents acid and ingested food from backing up and the pylorus sphincter muscle prevents food from leaving the stomach prematurely.

Each ball of food, or bolus, enters the part of the stomach called the fundus and pushes food previously eaten down and out towards the stomach walls. The fundus and the main body of the stomach serve as a storage area, holding food until it is time for it to move along to the antrum and the duodenum pronounced "du-o-dee-num".

The stomach has three coats or coverings;

1. The outer coat of serous membrane
2. The middle muscular coat
3. The inner mucous membrane

The inner mucous membrane is arranged in folds or rugae, which disappear when the stomach is distended. The membrane is lined with glands which produce gastric juice. This contains the enzymes pepsin (responsible for protein digestion) and rennin (responsible for the curdling of milk) as well as hydrochloric acid.

The middle layer contains a network of blood vessels that nourish the structure, as well as nerves that activate glands and muscles.

The outer or third layer are the muscles that move food, and in certain parts of the system, soften it and mix it with chemicals. This last layer is also a protective covering.

The hydrochloric acid the stomach secretes is corrosive enough to dissolve a razor blade or annihilate living cells. Sometimes it actually eats into the stomach itself and creates ulcers. Usually, however, the stomach remains impervious to attack.

First, the gastric lining is coated with mucus, which forms a barrier between the acid and the stomach and the stomach walls.

The mucus, somewhat alkaline, neutralises the acid and thus helps to keep the stomach from digesting itself. Furthermore, the stomach lining sheds cells at the rate of half a million a minute and replaces them so rapidly that the stomach has what amounts to a new lining every three days. Even if hydrochloric acid does damage the cells, the stomach makes repairs automatically.

The Small Intestine

From the stomach the food passes into the small intestine, the first part of this being the duodenum which is about ten inches long and shaped like a letter "c". As food enters the duodenum from the stomach, it stimulates four different organs to release the chemicals needed to finish digestion.
They are:
1. The small intestine pours forth mucus to protect the duodenum from damage by gastric acid. It also produces hormones that stimulate the liver, pancreas and gall bladder to release digestive substances.
2. The gallbladder stores and releases bile.
3. The pancreas produces an alkaline juice to neutralize acid.
4. The liver, described later in this lesson.

The remainder of the small intestine comprises the "jejunum", pronounced " je-joo-num", which is about eight feet long and the ilium which is about twelve feet long. The inner coat of the small intestine is comprised of mucous membrane arranged in folds known as valvulae conniventes, and unlike the rugae of the stomach these folds do not disappear with the distension of the intestines.

The mucous membrane is covered with minute finger-like projections known as villi, each villi contains a lacteal for the absorption of fat, and a capillary loop for the absorption of sugar and protein.

This mucous membrane also contains intestinal glands which produce a secretion known as succus entericus which contains enzymes for the digestion of proteins and sugars. The mucous membrane is studded with lymphatic nodes and in the latter part of the small intestine, which is known as the ilium, groups of these nodules are found and are known as peyers patches.

The function of these peyers patches is to fight infection. The small intestine then merges with the large intestine which although wider than the small intestine, is much shorter, being in total, 5-6 feet long.

The Large Intestine

For the purpose of this description, the large intestine is divided into nine parts. Starting with the caecum into which the ilium opens, the opening being guarded by the ileo-caecal valve which allows on-flow but prevents back-flow of intestinal contents. Then we have the vermiform appendix, which is about three inches in length and terminates in a blind end. Next, there is the ascending colon which passes up the right side of the abdomen to bend sharply at the right hepatic flexure to the transverse colon, the left splenic flexure and the descending colon which goes down the left side of the abdomen to the sigmoid colon (flexure); and the rectum which is five to six inches long, the exit of which, guarded by two sphincter muscles, known as the anus.

DIAGRAM OF THE LARGE INTESTINE

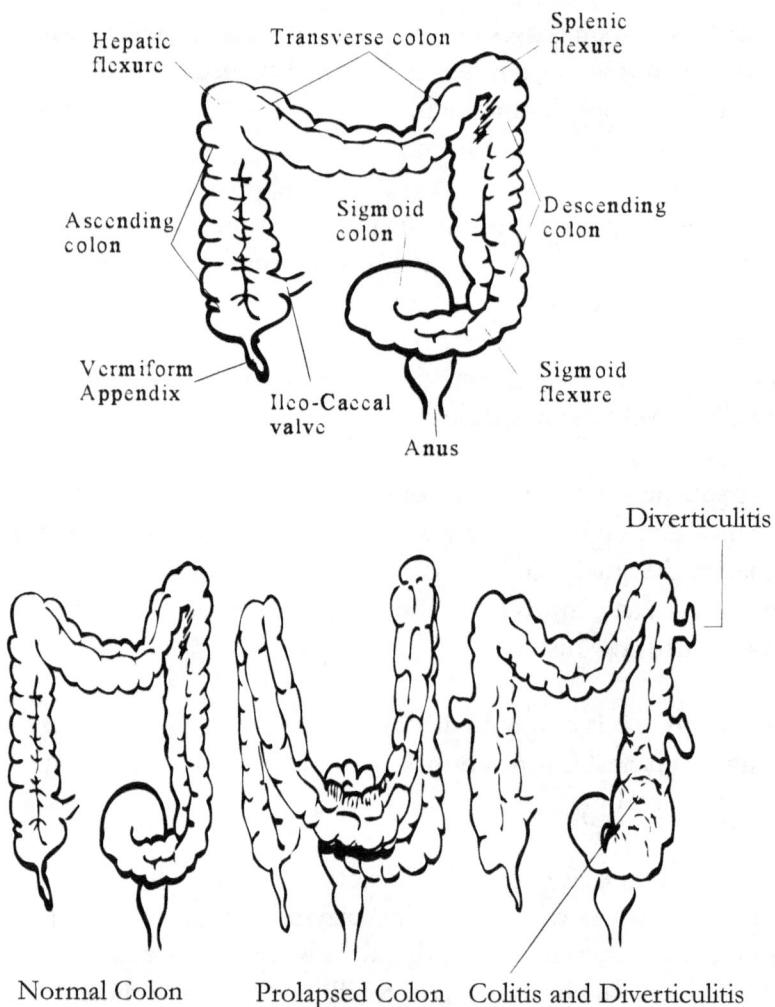

Normal Colon   Prolapsed Colon   Colitis and Diverticulitis

Having examined the alimentary canal it is necessary to look at the supporting organs of digestion.

The Liver

The liver is situated in the right hand side of the body just below the diaphragm. This is really a gland and is the largest organ in the body. It measures about ten to twelve inches across and six to seven inches from back to front, weighing approximately three and a half pounds. Protected by the rib cage, it is divided into two lobes, the large right lobe and the smaller left lobe overlaying the stomach where it joins the oesophagus.

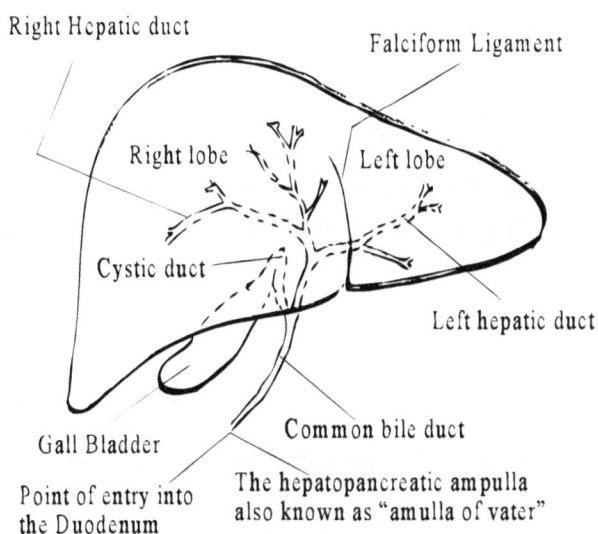

Right Hepatic duct

Falciform Ligament

Right lobe

Left lobe

Cystic duct

Left hepatic duct

Gall Bladder

Common bile duct

Point of entry into the Duodenum

The hepatopancreatic ampulla also known as "amulla of vater"

DIAGRAM OF THE LIVER

The right lobe is much larger than the left, and is subdivided into three sections. The liver can function even if as much as 90% of it was removed. If the liver were destroyed by disease, the only hope for survival would be a liver transplant. Because of its many functions, it is unlikely that a machine could do all of its work. The liver is one of the few organs capable of repairing itself to be fully functional again, given time and a change in lifestyle.

The liver has many functions, one of these is the formation of bile, which it produces up to two pints a day. This passes along to the gall bladder which is a muscular pear shaped sac about three inches long. Its function is to store bile and to concentrate it, which it does by eight to ten times, then as and when required, the bile passes out of the gall bladder into the duodenum. The Vitamins A, B, E, and K are stored in the liver. All but the "B" vitamins are soluble in fats and are therefore absorbed into the body with fatty materials.

The detergent action of bile in the small intestine breaks down these fats, along with the vitamins, into suspended globules that are absorbed through the intestinal wall and get into the blood via lymph vessels. Without bile, the body would be deficient in its vital stockpile of vitamins.

It is worth noting that a healthy liver removes a yellow pigment called "bilirubin" from the blood, converts it to a form which can be excreted into bile, and eliminates it from the body. A diseased liver, however, cannot do that, so the pigment remains in the bloodstream, and the skin and the whites of the eyes take on the yellowish tinge called "jaundice". It is interesting to note that when people talk about "yellow jaundice", they are repeating themselves, because the term "jaundice" is derived from the word that means "yellow".

Bilirubin is a waste product from the destruction of worn-out red blood cells. Under normal conditions, it gives the stool its characteristic colour. When a person has jaundice, the urine and the tears darken, but the stool becomes lighter.

It has already been seen that the liver manufactures bile but it has a variety of other functions. It is a powerful detoxifying organ, breaking down many kinds of toxic molecules and rendering them harmless. It is a reservoir for blood, a storage organ for some vitamins and particularly for digested carbohydrates in the form of glycogen which it releases to sustain blood and sugar levels. It manufactures enzymes, cholesterol, proteins, vitamin "A" from carotene, blood coagulation factors and other elements.

Bile is a complex fluid containing amongst other things bile salts and bile pigments. The pigment is derived from the disintegration of red blood cells and it is these which give the yellow brown colour to faeces which is excreted, while the bile salts are reabsorbed and reused. The salts promote efficient digestion of fats by a detergent action which gives very fine emulsification of fatty materials.

The Pancreas

The Greek name pancreas, means *"all flesh"* or *"all meat"* the pancreas is a cream coloured gland, 6-8 inches long and about 1 ½ inches wide. Resembling a fish, with a large head and a long tail, the pancreas extends across the body, behind the stomach, in the upper left side of the abdomen. The larger end of it rests next to the duodenum. **The function of the pancreas is to secrete enzymes and hormones, including insulin, that are needed for the digestion and absorption of food.** Insulin is manufactured by cells known as the **islets of langerhans**, which are scattered like little islands throughout the pancreas.

The pancreas is sometimes described as two organs in one:
- the **exocrine cells** of the pancreas secrete digestive enzymes into the duodenum
- the **endocrine cells** release two hormones, **glucagon and insulin**, into the blood

Glucagon acts in exactly the opposite way to insulin. While both hormones govern the level of glucose in the blood stream, glucagon is secreted in response to glucose deficiency. If there is insufficient response by this hormone, the result is hypoglycaemia.

Insulin regulates the utilization of glucose in the body. All body tissues except the brain require insulin for the absorption of glucose. If the pancreas fails to produce insulin, or secretes it in insufficient quantities, the result is a serious disease called diabetes **mellitus**.

A duct running the length of the organ, collects pancreatic juice and passes it to the duodenum at the same point that the **common bile duct** passes in bile.

The islets of langerhans are specialised cells of the pancreas which produce insulin. This passed into the general circulation and **controls carbohydrate metabolism**.

During the process of digestion, the larger particles of protein, carbohydrates, and fats must be reduced in size and converted into simpler substances to enable them to be absorbed through the walls of the digestive tract and into the blood stream. **Proteins are converted to peptones and polypeptides** which are then converted into amino acids.

Similarly, carbohydrates must be reduced, the large particles, **starches or polysaccharides**, are reduced to disaccharides which in turn are reduced to **monosacharides**. Fats are split into their component parts, fatty acids and glycerine. It should be noted that with one or two exceptions, there is no absorption of food elements until they reach the intestine where fatty acids and glycerine pass into the lacteals of the villi and amino acids into the capillary blood vessels. Fatty products are conveyed into the lymphatic system entering the systemic circulation via the thoracic duct. Amino acids and simple sugars are carried by the portal veins to the liver.

The movement of food along the digestive tract is made possible by wave-like muscular contractions known as **peristalsis**, the action is from the outside of the digestive tubes inwards and downwards, so that the food is forced further along the tube.

The stomach, being a muscularly controlled sac, is always on the move. It might be compared with an old fashioned milk churn where the food is pushed around and around until it is well and truly mixed with gastric juice which is a mixture of enzymes in **hydrochloric acid**.

As we have already seen, the stomach has a **pyloric valve** at the point where it merges with the small intestine. The function of this valve is to control the releases of the partially digested food material into the small intestine. Watery foods, such as soup, leave the stomach quite quickly whilst fats remain in the stomach longer. On average, an ordinary mixed diet meal is emptied from the stomach in 3-6 hours.

Some Conditions of the Digestive System

Disease or Disorder	**APPENDICITIS**
Description	This is an acute inflammation of the vermiform appendix. A distended and inflamed appendix may rupture, in which case it produces toxic materials and can cause peritonitis which is an acute inflammation of the abdomen. The inflammation is caused by an obstruction such as a hard mass of faeces or a foreign body in the lumen of the appendix, fibrous disease of the bowel wall, an adhesion, or a parasitic infestation. The most common symptom is constant pain in the lower right quadrant of the abdomen. The patient often feels intermittent pain in the mid-abdomen and gets relief by bending the knees and waist into the fetal position. Vomiting and a low-grade fever of 99degs.-102degs., an elevated white blood count, rebound tenderness, and a rigid abdomen are accompanying symptoms.
Possible Allopathic Treatment	Appendectomy, within 24 - 48 hours of the first symptoms, delay may result in rupture and peritonitis as faecal matter is released. The fever will rise sharply once peritonitis begins.

Disease or Disorder	**PERITONITIS**
Description	An inflammation of the peritoneum often produced by either bacteria or an irritating substance introduced into the abdominal cavity via a penetrating wound or perforation or rupture of an organ in the Gastro-Intestinal tract or the reproductive tract. The most common cause is a rupture of the veriform appendix or perforations in the Colon due to diverticulitis. Secondary causes are peptic ulcers, gangrenous gallbladder, gangrenous obstructions of the small bowel, ruptures of the spleen, liver, ovarian cyst, or fallopian tube.
Possible Allopathic Treatment	The patient is hospitalised and placed in a bed in a semi-Fowler's position with knees flexed to facilitate breathing and localize pus in the lower abdomen. Oxygen, parenteral fluids with electrolytes, antibiotics and emetics are administered as ordered. Surgery is usually delayed until the patient is stabilized. Patients who respond to antibiotics receive a liquid diet.

Disease or Disorder	**CIRRHOSIS OF THE LIVER**
Description	Cirrhosis of the liver is the breakdown of normal liver tissue that leaves non-functioning scar tissue surrounding areas of functioning liver tissue. There are several types of cirrhosis of the liver but portal cirrhosis is by far the most common. This is also referred to gin drinkers liver, or alcoholic liver. It is usually caused by exposure to poison which can include such substances as carbon tetrachloride and phosphorous but, by far the most common cause is the ingestion of alcohol. This makes the liver leathery and produces on its normally smooth surface, nodules varying in size from a pin head to a bean which gives it a hobnailed appearance. Other common causes are nutritional deprivation and hepatitis. The symptoms are nausea, fatigue, anorexia, weight loss, ascites, varicosities and spider angiomas.
Possible Allopathic Treatment	Treatment depends on the etiology. The liver has remarkable regenerator abilities, but recovery is slow.

Disease or Disorder	**JAUNDICE**
Description	Jaundice is normally evidenced by the yellowness of the skin, mucus membranes and sclerae of the eyes, caused by an excess of bile pigments known as bilirubin in the circulatory system. It may occur when the outflow of the bile has been blocked and when the liver becomes obstructed. A large portion of the bile which is produced by the liver is absorbed directly into the blood stream, as it cannot flow normally out of the bile duct into the duodenum. Persons with jaundice may also experience nausea, vomiting, abdominal pain and dark urine. Jaundice is a symptom of many disorders related to the liver.
Possible Allopathic Treatment	Useful diagnostic procedures include a clinical evaluation of the signs & symptoms, tests of liver function, and techniques for direct or indirect visualization ie. X-Ray, Cat scan, Ultrasound, endoscopy or exploratory surgery and biopsy.

Disease or Disorder	**ACUTE PANCREATITIS**
Description	If the pancreas protein-digesting enzymes are activated while in the pancreas, they are powerful enough to digest the pancreas itself. This condition, known as acute pancreatitis, can occur if the pancreatic duct is obstructed and the digestive enzymes are forced to accumulate in the pancreas. When this happens, the substances that normally inhibit the activation of the enzymes are overwhelmed, and the pancreas may be damaged or even destroyed by its own juices. Acute pancreatitis is generally the result of damage to the biliary tract, such as by alcohol, trauma, infectious disease, or certain drugs. Characterized by severe abdominal pain radiating to the back, fever, anorexia, nausea and vomiting.
Possible Allopathic Treatment	Treatment includes nasogastric suction to remove gastric secretions. To prevent stimulation of the pancreas, nothing is given by mouth. Intravenous fluids and electrolytes are administered and nonmorphine derivatives are given to relieve pain.

Disease or Disorder	**HEARTBURN (pyrosis)**
Description	This is a common symptom of gastric distress consisting of a burning sensation which extends up into the oesophagus from the base of the sternum and is quite often felt to rise into the throat and may be accompanied by a sour belch. Heartburn is usually caused by the reflux of gastric contents into the esophagus but may be caused by gastric hyper-acidity or peptic ulcer.
Possible Allopathic Treatment	Anti-acids relieve the symptoms but do not cure the heartburn.

Disease or Disorder	**ULCERATIVE COLITIS**
Description	Colitis is an inflammatory condition of the large intestine, with symptoms of diarrhea, bleeding and ulceration of the mucosa of the intestine, weight loss and pain. Ulcerative colitis is a chronic episode of colitis with increased symptomology.
Possible Allopathic Treatment	Diagnosis is based on clinical signs, barium X-Ray of the colon and colonoscopy with biopsy. Steroids, fluids, electrolytes, antibiotics are used with colitis and chronic ulcerative colitis may be treated with corticosteroids and anti-inflammatory agents. Surgery may be an option if drug therapy is ineffective

Disease or Disorder	**HEPATITIS**
Description	Hepatitis, or inflammation of the liver, characterized by jaundice, hepatomegaly, anorexia, abdominal and gastric discomfort, abnormal liver function, clay-coloured stools, and tea-coloured urine. Hepatitis is sometimes caused by alcohol or by certain drugs or chemicals, bacterial or viral infection, parasitic infestation, but in many instances it is the result of one kind of three viruses, known as type A, type B and non-A, non-B. That awkward designation for the third virus comes from the fact that the specific agents have yet to be identified, doctors only know that it isn't either A or B.
	Hepatitis 'C' (non-A, non-B) A type of hepatitis transmitted largely by blood.
Possible Allopathic Treatment	Regardless of type, viral hepatitis results in a common set of symptoms, which include fever, headache, sore throat, nausea, aching joints and muscles, loss of appetite, weakness, pain in the upper right abdomen, and jaundice.
	NOTE: Do not massage any person with a contagious disease. This disease can still be contracted after many years from the recovery date.

Disease or Disorder	**HEPATITIS A**
Description	Primarily a disease of young adults or children is spread through faecally contaminated food, water or contaminated objects. Although victims may feel miserable, the illness almost never has lasting consequences. A form of infectious viral hepatitis caused by the hepatitis "A" virus. Also called acute infective hepatitis. Prophylaxis with immune globulin is effective in household and sexual contacts.
Possible Allopathic Treatment	Regardless of type, viral hepatitis results in a common set of symptoms, which include fever, headache, sore throat, nausea, aching joints and muscles, loss of appetite, weakness, pain in the upper right abdomen, and jaundice.
	NOTE: Do not massage any person with a contagious disease. This disease can still be contracted after many years from the recovery date.

Disease or Disorder	**HEPATITIS B**
Description	A form of viral hepatitis caused by the hepatitis B virus. The virus is transmitted in contaminated serum in blood transfusion, sexual contact with an infected person, or by contaminated needles and instruments. The infection may be severe and result in prolonged illness, destruction of liver cells, cirrhosis, or death. On the other hand is caused by chronic liver inflammation in about 10% of all patients and can be serious for the elderly or those in poor health. Serum hepatitis, as this ailment used to be called, is transmitted through blood transfusions, injections with unsterile needles, or intimate contact (body fluids, saliva, semen or tears).
Possible Allopathic Treatment	Regardless of type, viral hepatitis results in a common set of symptoms, which include fever, headache, sore throat, nausea, aching joints and muscles, loss of appetite, weakness, pain in the upper right abdomen, and jaundice.

NOTE:
Do not massage any person with a contagious disease. This disease can still be contracted after many years from the recovery date.

Disease or Disorder	**DIVERTICULITIS**
Description	Inflammation of small pouches, called "diverticula" on the colon. If faecal matter penetrates through the thin wall of the diverticula it will cause inflammation and possible abscess formation in the tissues surrounding the colon. During this period the lumen of the colon often narrows and may become obstructed. The patient will experience crampy pain, particularly over the area of the sigmoid colon, fever, and leucocytosis are also present.
Possible Allopathic Treatment	Barium enemas and proctoscopy are performed to rule out carcinoma of the colon, (which exhibits many of the same symptoms). Non-invasive treatment may include, bed rest, intravenous fluids, antibiotics and nothing taken by mouth. In acute cases, bowel resection of the affected part greatly reduces mortality and morbidity.

Disease or Disorder	**DUODENAL ULCER**
Description	More commonly referred to as peptic ulcer. Ulceration of the stomach or first part of the duodenum from excessive acid in the stomach or inadequate resistance to acids of the mucosal lining. Ulcers may be acute or chronic. Acute lesions are almost always multiple and superficial. Chronic ulcers are deep, single, persistent and symptomatic. The muscular coating of the stomach wall does not regenerate; during healing a scar will form permanently marking the site. Ulcers are caused by a variety of poorly understood or misunderstood factors, that include, excessive secretion of gastric acid, inadequate protection of the mucus membrane, stress, heredity, and the taking of certain drugs including corticosteroids, anti-hypertensive and anti-inflammatory medication.
Possible Allopathic Treatment	Short term symptomatic relief is provided with antacids and frequent, small, simple meals. Haemorrhage may be caused by perforation of the muscle and blood vessels and often requires surgical resection of the damaged area. The diagnosis and evaluation of peptic ulcers involve serial X-rays using a contrast medium and endoscopy.

Disease or Disorder	**GASTRITIS**
Description	Acute inflammation of the stomach that may be due to chemical irritation or infection that occurs in two forms; acute and chronic. Acute gastritis, may be caused by severe burns, major surgery, aspirin or other anti-inflammatory agents, corticosteroids, drugs, food allergens, viral, bacterial, or chemical toxins. The symptoms of anorexia, nausea, vomiting and discomfort after eating, usually abate after the causative agent has been removed. Chronic gastritis may occur if the irritation is continued for a long period, and is usually a sign of an underlying disease, such as peptic ulcer, stomach cancer, severe peptic ulceration (zollinger-Ellison syndrome), or pernicious anaemia. Differential diagnosis is by endoscopy with biopsy.
Possible Allopathic Treatment	If torrential bleeding occurs the reported mortality rate is 60%. Surgical intervention with low recovery effects. Anti-secretory ulcer drugs, vasoconstrictors and coagulation medication have proven useful. Acute onset is a happier picture, prevention and/or early treatment before bleeding is a problem will increase the probabilityof recovery.

Disease or Disorder	**HAEMORRHOIDS**
Description	Venous distension in the anal area causing dilation of the bloodvessels and swelling of the mucosa overlying them. Internalhaemorrhoids originate above the internal sphincter of the anus. Ifthey become large enough to protrude from the anus, they maybecome constricted and painful. Small internal haemorrhoids often bleed with defecation. External haemorrhoids appear outside the anal sphincter. They are usually not painful, and bleeding does not occur unless a haemorrhoidal vein ruptures or thrombosis.
Possible Allopathic Treatment	Treatment includes local applications of a topical medication to lubricate, anaesthetize, and shrink the haemorrhoids, sitz baths and cold or hot compresses are also soothing. As a last resort surgery may be necessary.

Disease or Disorder	**MUMPS**
Description	Virus infection of the salivary glands that may also involve the pancreas. Swelling of the parotid glands caused by paramyxovirus. Normally it effects children between 5-15 years of age. Adults that are infected react severely. The mumps paramyxovisus lives in the saliva of the effected individuals and is transmitted in droplets or by direct contact. The virus is present in the saliva from 6 days before to 9 days after the onset of the swelling of the parotid glands. The disease sometimes involves complications, such as arthritis, pancreatitis, myocarditis, oophoritis, and nephritis. About one half of the men with mumps suffer some atrophy of the testicles, sterility rarely results. Symptoms include, anorexia, headache, malaise, low grade fever, earache, parotid gland swelling and a temperature of 101 - 104 degrees F.
Possible Allopathic Treatment	Respiratory isolation and analgesics, antipyretics and plenty of fluids.

Terminology

Ampulla of vater: Point at which the common bile and pancreatic ducts enter the duodenum.

Amylase: Enzyme that converts starch into maltose.

Anal sphincter: Circular muscle at the end of the digestive tract.

Appendix: Small finger-like protuberance from the caecum.

Bile duct: Conveys bile from the cystic and hepatic ducts to the bile duct and into the duodenum at the ampulla of vater.

Bile-pigments: Dark-coloured substance formed by the breakdown of red blood cells.

Bile salts: Complex salts excreted by the liver that help in the emulsification of intestinal fats.

Bilirubin: Principal bile pigment.

Enterogastrone: Hormone secreted by the duodenum to slow down the gastric action.

Erepsin: Duodenal enzyme to break peptones into amino-acids.

Gastrin: Hormone produced by the stomach to maintain the gastric secretions.

Glycogen: Simple form of starch that can be broken down into glucose.

Glycogon: Hormone secreted in response to low glucose level of blood.

Ileo-caecal valve: Valve that allows the chyme to pass from the terminal ileum into the caecum but not return.

Insulin: Glucose-controlling hormone secreted by the pancreas.

Maltase: Duodenal enzyme to reduce sugar to glucose.

Pepsin: Gastric enzyme that splits protein into peptones.

Rennin: Gastric enzyme that curdles milk.

Trypsin: Pancreatic enzyme that splits peptones and proteins into amino-acids.

Urea: Nitrogen containing substance formed from ammonia by the liver.

The Vascular System

Introduction

This system which is sometimes called the circulatory system, consists of the heart, blood vessels, blood, lymphatic vessels and lymph.

The centre of the circulatory system, is the heart, which is a muscular organ that rhythmically contracts, forcing the blood through a system of vessels. The heart itself weighs approximately nine ounces (225 grammes) in a fully grown adult and lies one-third to the right and two-thirds to the left of the thoracic cavity. At birth it beats about 130 times a minute, at six years about 100 times a minute, reducing in adult life to between 65-80 beats a minute, giving an average of about 74.

During the 24 hour period, an adult human heart will pump about 9,000 litres of blood through approximately 12,000 miles (20,000 Klm) of blood vessels.

The Heart

The heart is encapsulated in a fibrous pericardium containing the serous pericardial sac, holding a small amount of fluid that allows frictionless movement.

The heart is divided into four chambers. These are the right and left atrium (or auricles). In the upper part of the heart, and the right and left ventricles in the lower part of the heart. The right side of the heart is divided vertically by a solid wall or septum, which prevents the venous blood in the right side from coming into contact or mixing with the arterial blood in the left side of the heart. The right side pumps deoxygenated blood through to the lungs, and the left side pumps oxygenated blood from the lungs. This is known as the systemic circulation (pulmonary circulation).

To keep the heart beating the Sino-Atrial Node located in the right atrium sends impulses through the two atria, causing atrial systole. It then stimulates the atrio-ventricular node to pass rapidly down the Bundle of His to cause ventricular systole.

Circulation is divided into two principle systems, the general or systemic circulation and the pulmonary circulation. The systemic system has two particular branches - the portal and the coronary circulation - but for the purposes of this text we will confine ourselves to the two principle systems: the systemic and the pulmonary.

THE HEART

Aorta

Pulmonary Artery

Superior
Vena Cava

Left Atrium

Pulmonary
Veins

Right Atrium

Left Ventricle

Inferior
Vena Cava

Right Ventricle

Septum

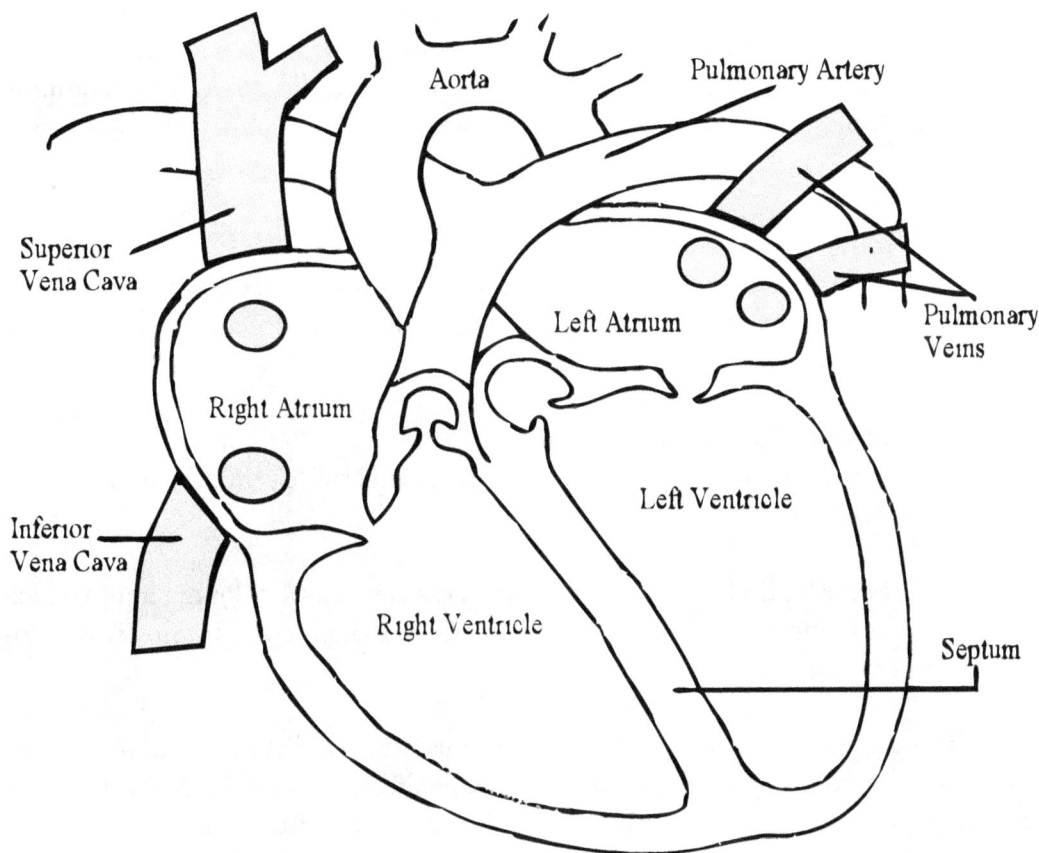

Blood vessels which proceed from the heart are known as arteries. These are hollow, elastic tubes which start off fairly large and gradually decrease in diameter as they spread throughout the body until they become known as arterioles and finally very fine hairlike blood vessels known as capillaries. The blood vessels which proceed towards the heart are known as veins, these are elastic tubes which have one way valves to prevent a backwards flow of the blood.

The veins empty their contents through the inferior and superior venacava into the right hand side of the heart, first of all into the right atrium, from there it passes into the right ventricle. Then on through to the pulmonary artery into the lungs where it is re-oxygenated and then returned to the heart via the pulmonary veins into the left atrium and then into the left ventricle until finally, into the body, through the ascending and descending aortas.

The descending aorta divides into two iliacs. When the iliac reaches the groin it then becomes the femoral, passing underneath the knee when it becomes the popliteal. It further sub-divides into the anterior tibial and the posterior tibial, then to the dorsalis pedis and the plantar arches. The ascending aorta divides into two subclavian which then pass underneath the arm.

In the armpit it becomes the axillary then the brachial for the length of the upper arm when it divides into the radial and ulna culminating in the palmer arches. When it reaches the subclavian junction another branch goes upwards through the common carotid to the facial, temporal and occipital branches.

Blood

Blood is alkaline in pH. Its quantity amounts to something like six litres in the average adult. It is very complex in nature but has four principle constituent parts:

1. Plasma.
2. Erythrocytes (red blood corpuscles).
3. Leucocytes (white corpuscles). Corpuscles is Latin for little bodies.
4. Platelets.

Plasma provides the liquid basis of the blood, this is a clear straw coloured liquid which holds various substances in solution. These include: water, glucose, amino acids, mineral salts and enzymes, etc.

Erythrocytes are inert biconcave discs, that are made in the red bone marrow. They get their colour from haemoglobin which has the ability to absorb oxygen (when it becomes oxy-haemoglobin, which is bright red in colour) and carbon dioxide when it becomes carboxi-haemoglobin which becomes very dark red, bordering on a muddy brown colour). Their function is to carry oxygen from the lungs to the tissues and carbon dioxide from the tissues to the lungs during respiration. The average life span of an erythrocyte is 120 days. As they age, they lose their elasticity and are trapped in the small blood vessels, spleen and other organs. Their eventual disintegration takes place in the spleen, being finally completed in the liver.

THE CIRCULATION SYSTEM

The Arteries

Temporal

Facial

Carotid

Subclavian

Axillary

Brachial

Aorta

Radial

Ulnar

Palmar Arches

Iliac

Femoral

Pophical

Posterior Tibial

Anterior Tibial

Plantar Arches

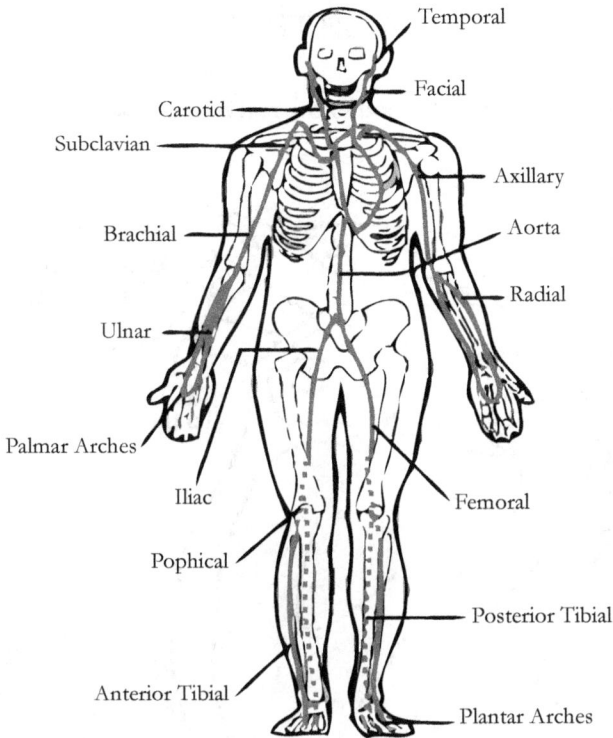

Direction of circulation

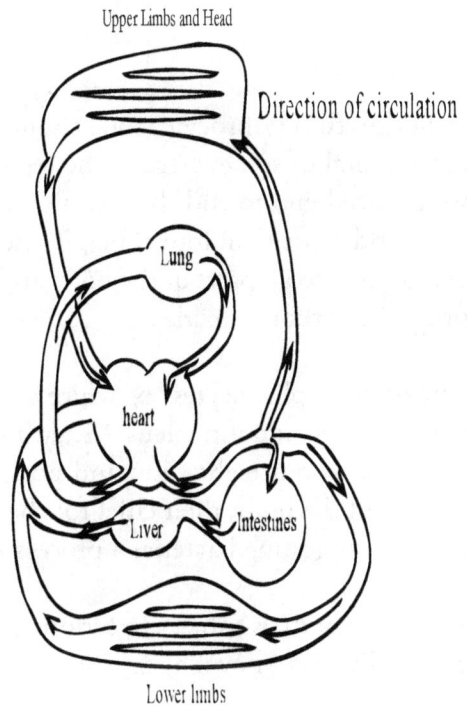

Upper Limbs and Head

Direction of circulation

Lung

heart

Liver

Intestines

Lower limbs

The following diagram shows the circulation of blood through the heart. This diagram is not drawn
accurately. But this method of drawing makes it easier to understand.

THE CIRCULATION OF BLOOD THROUGH THE HEART

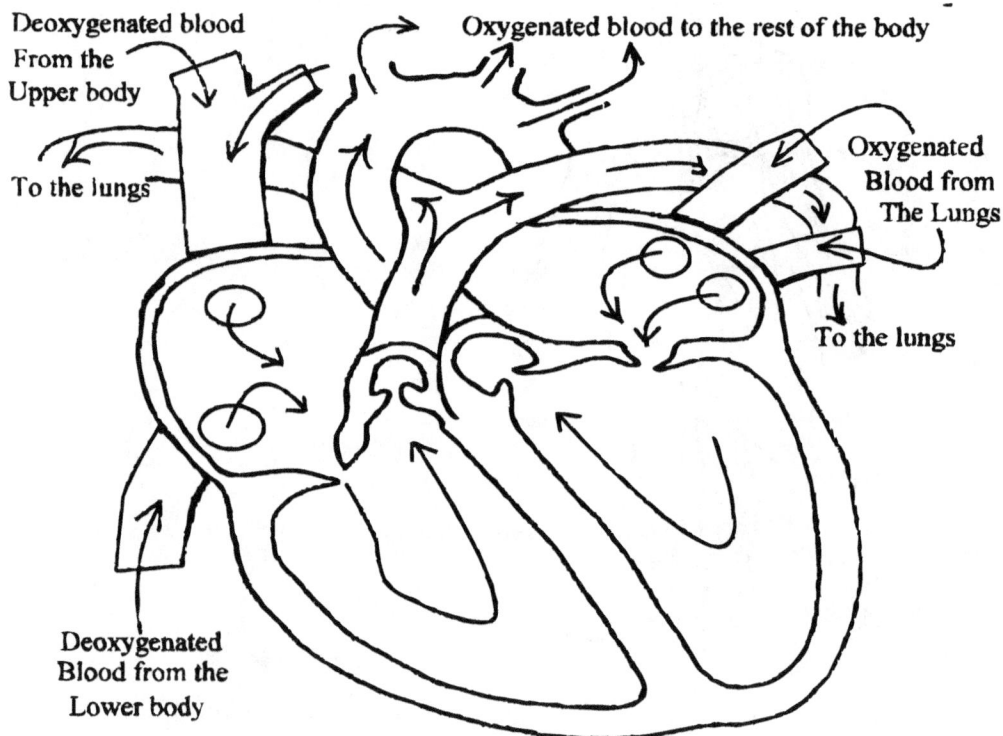

Deoxygenated blood
From the
Upper body

Oxygenated blood to the rest of the body

To the lungs

Oxygenated
Blood from
The Lungs

To the lungs

Deoxygenated
Blood from the
Lower body

In health, the erythrocytes total about five million per cubic millimetre of blood, which gives a total of somewhere in the region of 25 billion in a human adult. If these cells were placed end to end they would form a ribbon sufficiently long enough to encircle the world more than four times. These cells are the body's transporters, they carry food and oxygen to all parts of the body and on their return journey pick up waste products, primarily carbon-dioxide.

Leucocytes (phagocytes) or white corpuscles are larger than erythrocytes, have an irregular shape and a nucleus. They are produced in the bone marrow and, in health, they total about eight thousand per cubic millimetre. They act as the protectors or soldiers of the body, their chief role being to protect the body against infection by their power of ingesting bacteria: a process which is known as phagocytosis.

When the body is subject to serious infection the leucocytes increase rapidly by a process of division known as mitosis.

Then we have platelets or thrombocytes. These average 250 thousand per cubic millimetre of blood, they are derived from large multinucleated cells in the bone marrow and are essential to the blood for coagulation, (clotting).

This diagram shows the open and closed vein valves:

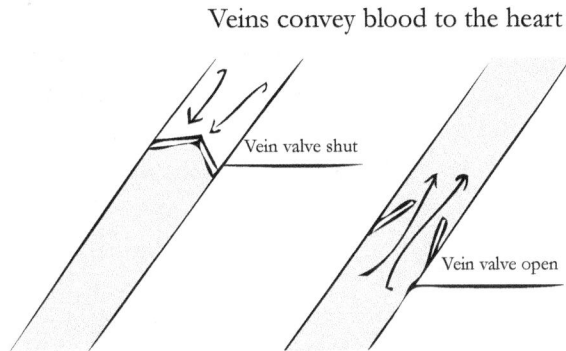

Veins convey blood to the heart

Vein valve shut

Vein valve open

BLOOD TYPES

The existence of human blood types was established by Karl Landsteiner in 1902, when he began a study to determine why fatalities occurred following some blood transfusions. He discovered that the cause was incompatibility between the blood of the donor and the blood of the recipient. Arising from this work came the Landsteiner classification of blood groups which classified blood into the four main blood types: "A;", "B", "AB", and "O".

Type "O" is sometimes called the universal donor because it may give blood to all blood types, but it can only receive from type "O".

On the other hand, type "AB" is sometimes called the universal recipient because it can receive from any group but can only give to the "AB" group.

Type "A" can give to both "A" and "AB" and receive only from blood types "A" and "O". Type "B" can give to type "B" and "AB" and receive only from blood types "B" or "O".

In 1940 Landsteiner and A.S. Weiner, recognized the "Rh" (Rhesus) factor, a substance found in red blood cells. This was discovered during their experiments with Rhesus monkeys, hence the name Rhesus or an abbreviation for Rhesus (Rh). It is estimated that 85% of white people have a Rh positive factor and the other 15% are Rh negative. The presence or absence of antigen makes the individual either Rhesus positive or Rhesus negative. Antibodies are not found in Rhesus negative people.

Some Conditions of the Vascular System

Disease or Disorder SIMPLE ANAEMIA

Description This is probably the most common blood complaint. Anaemia means, loss of normal balance between the productive and destructive blood processes. This can be due to a drop in the blood volume after a haemorrhage, or the drop in the number of red blood cells, or in the amount of haemoglobin, or a combination of any two or more of these factors. There are many forms of anaemia, but we are primarily concerned with two categories:

1. simple anaemia. 2. pernicious anaemia.

In simple anaemia there are two direct causative factors, the first is a marked nutritional deficiency of iron, frequently seen with pregnant women, and premature infants, as well as in growing children. The second causative factor is through chronic blood loss, during menstruation or because of an accident.

One of the characteristics of pernicious anaemia is the presence of giant red cells (Macrocytes), each cell appearing to be overloaded with haemoglobin, while the total red cell count is decreased. As recently as 1925 this disease was invariably fatal, today the life expectancy of the properly treated patient is no different from that of the general population. Pernicious anaemia results from failure of the red blood cells to develop and mature normally. While a decreased number of red blood cells is indicative of anaemia a continuously increasing number of white blood cells can be indicative of leukaemia. Reference has already been made to the fact that white blood cells

increase in number by mitosis, in the presence of necessary stimuli such as an infection and the normal eight thousand per cubic millimetre of blood can increase to as many as sixty thousand, in a case of severe pneumonia. However, when the condition is cured, the mitosing effect ceases and the white blood count returns to normal. In leukaemia the leucocytes and/or lymphocytes do not remain at the normal number, but gradually increase.

Possible Allopathic Treatment The therapeutic response to anaemia depends on the cause. Moderate to severe anaemia may require a blood transfusion. If the condition is acute, a supplement of the deficient component will be administered

Disease or Disorder	**BRUISES**
Description	A bruise could be described as a discolouration in or below the skin tissue, when tiny blood vessels or capillaries are ruptured. Blood seeps into tissues and the red cells break. Discolouration disappears as the red cells degenerate and are reabsorbed, a bruise disappears usually in about 14 days.

A blue bruise appears if the overlying tissue is pink. Haemoglobin the principle pigment in red cells turns blue when oxygen is removed. Tissues quickly use up oxygen and it is not resupplied because the blood vessels are broken. A yellow bruise occurs due to the breakdown products of haemoglobin. The pigment is eventually removed by white blood cells. One of the reasons for the yellow colour is due to an iron deficiency. With certain blood disorders that affect the platelets there can be unusual bruising which can be caused with only very slight pressure on the skin.

Possible Allopathic Treatment

In the case of trauma to the head, (contusions) the patient should be closely observed for changes in neurological signs, including mental status, vital signs, pupillary findings, focalization or lateralizations of signs and seizures. Arnica cream will help relieve and cause the bruising to come to the surface quicker.

Disease or Disorder   **ARTERIOSCLEROSIS and ATHEROSCLEROSIS**

Description

Arteriosclerosis: Arteriosclerosis is a general term used to describe different conditions relating to hardening of the arteries. Some hardening will occur as we age. It can be caused by atherosclerosis.

Atherosclerosis: This is where a build up of fats, plaque (fibrous tissue, cholesterol and calcium) on the inside of the artery reducing the size of the wall.

Possible Allopathic Treatment

Diet and lifestyle changes including exercise are recommended. Vasodilators and exercise may relieve symptoms, but there is no specific treatment for the disorder. Surgery may be an option with chronic severe cases.

Disease or Disorder	**VARICOSE VEINS**
Description	A network of veins serves to drain the capillary beds and body tissue of "used" blood, and returns this blood to the heart. Venous flow is assisted in its return to the heart by the rhythmic suction action of breathing, the muscular contraction in the extremities and the valves located in the veins. Gravity assists the venous blood from the neck and head to return to the heart, but venous flow from the legs is against the pull of gravity, and for most of the day has to run up hill. The valves in the veins prevent back flow and when some of these valves become impaired or cease to function, the veins become permanently dilated. There are many causes of varicosity but these include;
	1. congenital factors; heredity
	2. environmental factors: people whose work necessitates standing still for long periods of time are at particular risk.
	Varicose veins are also quite often, a complication of pregnancy.
Possible Allopathic Treatment	The saphenous veins of the legs are most often affected. Elevation of the legs and the use of elastic stocking are often advised. Surgery (ligation & stripping) may be required. Injection of sclerosing solutions help prevent or treat postphlebitic syndrome.

Disease or Disorder	**HAEMOPHILIA**
Description	This is the best known of the bleeding diseases. It is an hereditary disease where the victim almost always is male, the disease being passed on by the mother, who is the carrier. It is a disease in which there is a deficiency in the clotting of the blood.
Possible Allopathic Treatment	The primary objective is to prevent bleeding and to make the environment as safe as possible.

Terminology

Angiology: The science of dealing with blood and lymph vessels.

Cholesterol: A major fat found in the blood and a constituent in animal fats and oils. There are different types of cholesterol found in the body. LDL (low density lipo-proteins) is considered a bad cholesterol and is present in many foods such as fast food and other modern day products. They increase the chances of hypertension and other cardio-vascular conditions. HDL (high density lipo-proteins) are considered good cholesterol as it helps remove excess cholesterol off the sides of the arteries.

Coronary: Usually relating to the blood.

Diastolic pressure: Is the pressure measured during the relaxing phase of the cardiac cycle.

Haemorrhoids (piles): Are dilated veins in the rectum and anus either internal or external.

Hypertension: A term relating to high blood pressure.

Hypotension: A term relating to low blood pressure.

Phlebitis: An inflammation of the vein walls, mostly found in the legs, which may lead to thrombo-phlebitis, which is a complication caused by an obstructing blood clot.

Systolic pressure: The pressure measured during the contraction phase of the cardiac cycle.

Thrombus: A clot of blood found within the heart or blood vessels.

Endocarditis: Inflammation of the heart lining, usually the valves.

Pericarditis: Inflammation of the pericardium.

Atrial fibrillation: Irregular, rapid heart beat due to disorder of the Sino-Atrial Node.

The Lymphatic System

Introduction

This is a type of **secondary circulation** inexplicably intertwined with the blood circulation. The basic material of the lymphatic system is the lymph which is **plasma** after it has exuded from the capillaries. It **gives nourishment** to the tissue cells and in return **takes away their waste** products. The liquid is drained off by tiny lymphatic vessels which join together to form larger lymph vessels and, as **these lymph vessels convey lymph towards the heart**, they are supplied with valves in much the same way as veins. Along their course towards the heart there are receiving or reservoir areas known as **Lymph Nodes**. They vary in size from pin head to a small almond.

The purpose of these lymph nodes is to filter the lymph and destroy foreign substances as it passes through and, in this way to help prevent infection passing into the blood stream and to produce and add lymphocytes to the lymph, and to produce antibodies.

Body cells live in tissue fluid, a liquid derived from the bloodstream. Water and dissolved substances, such as oxygen and nutrients, are constantly filtering through capillary walls into spaces between cells and as a result are adding volume to tissue fluids. Under normal conditions, fluid is also constantly removed so that it does not accumulate in the tissues. Part of this fluid simply returns (by diffusion) to the capillary bloodstream, taking with it some of the end products of cellular metabolism, including carbon dioxide and other substances. A second pathway for drainage of tissue fluid involves the lymphatic system. In addition to blood-carrying capillaries, there are microscopic vessels called lymphatic capillaries, which drain away excess tissue fluid that does not return to the blood capillaries. The relation between the circulatory system and the limbic system is shown below.

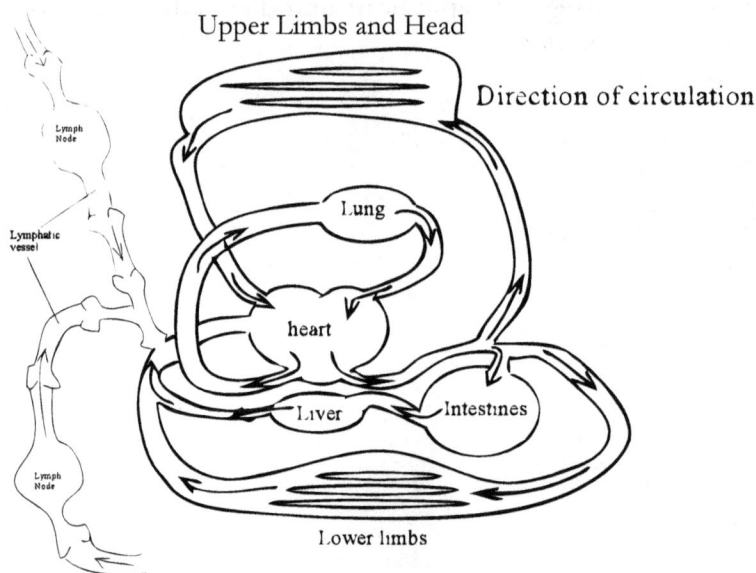

Upper Limbs and Head

Direction of circulation

Lymph Node

Lymphatic vessel

Lung

heart

Liver

Intestines

Lymph Node

Lower limbs

Another important function of the lymphatic system is to absorb protein from the tissue fluid and return it to the bloodstream. As soon as tissue fluids enters the lymphatic capillary, it is called "lymph". The lymphatic capillaries join to form the larger lymphatic vessels or ducts, and these vessels eventually empty into the veins. But, before the lymph reaches the veins, it flows through filters called "lymph nodes", this is where bacteria and other foreign particle are trapped and destroyed.

Lymphatic capillaries resemble blood capillaries in that they are made of one layer of flattened Squamous epithelial cells also called endothelium cells, which allows for easy passage of soluble material. Gaps between these endothelial cells allow the entrance of proteins and other relatively large suspended molecules. Unlike the blood stream the lymph seems to have no direction other than forwards and in the general direction of the heart.

The lymphatic vessels are thin walled and delicate and have a beaded appearance because of indentations where valves are located. These valves prevent backflow in the same way as veins by having a one-way valve positioned at intervals along the way.

Surface or superficial lymph is found within 2cm below the skin surface, often near vein pathways. The deeper lymph vessels are larger and accompany the deep veins.

Lymph vessels are named according to location. For example, lymph found in the breast area is called "mammary lymph", lymph vessels in the thigh are called "femoral lymph vessels". All lymph vessels carry lymph to selected areas called lymph nodes. Lymphatic vessels carry lymph away from the regional nodes to eventually drain into one of two terminals. The **right lymphatic duct** or the **left thoracic duct**, to be emptied into the bloodstream.

Lymphoid tissue or specialized lymph glands are:
1. **Tonsils**
2. **Adenoids**
3. **Peyer's patches in the ileum.**

Eventually all lymph passes into two principal lymph vessels, the thoracic duct and the right lymphatic duct, which opens into the blood stream at the junctions of the right and left, **internal jugular** and **subclavian veins** where it becomes part of the general systemic circulation again.

The right lymphatic duct is a short vessel about 1.25cm long that receives lymph that comes from the upper right quadrant of the body (*right side of head, neck, arm and thorax*). It empties into the **right subclavian** vein which is guarded by two **semilunar valves** to prevent blood from entering the duct. The rest of the body is drained by the **thoracic duct** which is much larger (*about 40cm long*).

The duct begins in the posterior part of the abdomen, just below the diaphragm. The first part of the duct is enlarged to form a cistern or temporary storage pouch, called the **cisterna chyli**. The word *chyle comes from the milky fluid* formed by the combination of fat globules and lymph that comes from the intestinal lacteals.

The thoracic duct extends upwards through the diaphragm and along the back wall of the thorax up into the root of the neck of the left side. Here it receives the left jugular lymphatic vessels from the head and neck, the left subclavian vessels from the left upper extremity, and other lymphatic vessels from the thorax. All lymph from below the diaphragm empties into the cisterna chyli by way of the various lymph nodes.

There are approximately 600-1000 of these lymphatic nodes scattered throughout the body along the line of the lymphatic vessels.

The most common superficial ones being:

1. **Cervical nodes** or profundus on either side of the neck just below the ears, are divided into deep and superficial groups which drain certain parts of the head and neck. They often become enlarged during upper respiratory infections.
2. **The Axillary glands** in the armpit. May become enlarged following infections of the upper extremities and breasts, cancer cells from the breasts often metastasize (spread) to the axillary nodes.
3. **Tracheobronchial nodes** are found near the trachea and around the larger bronchial tubes. People living in polluted areas, these nodes become filled with black carbon particle.
4. **Mesenteric nodes** are found between two layers of peritoneum that forms the mesentery (membrane around the intestines). There are about 100 - 150 of these nodes.
5. The **Inguinals** in the groin. Receives lymph from the lower extremities and genital organs. When they are enlarged they are often referred to as buboes from which the bubonic plague got its name.
6. The ones in the **Popliteal Fossa** or depression behind the knee.
7. The **Supratrochlea** or cubital nodes in the crutch of the elbow.
8. The **Supraclavicular** glands.
9. The **Submandibular nodes** underneath the mandible and the cervical, and occipital glands.

This is not a complete list of lymph nodes, but they are sometimes the most obvious of the superficial glands which swell when an infection is present in that part of the body.

Submandibular

Thoracic duct

Thoracic duct

Axillary glands

Axillary glands

Subratrochlea

Subratrochlea

Inguinals

Cisterna chyli

Popliteal

Popliteal

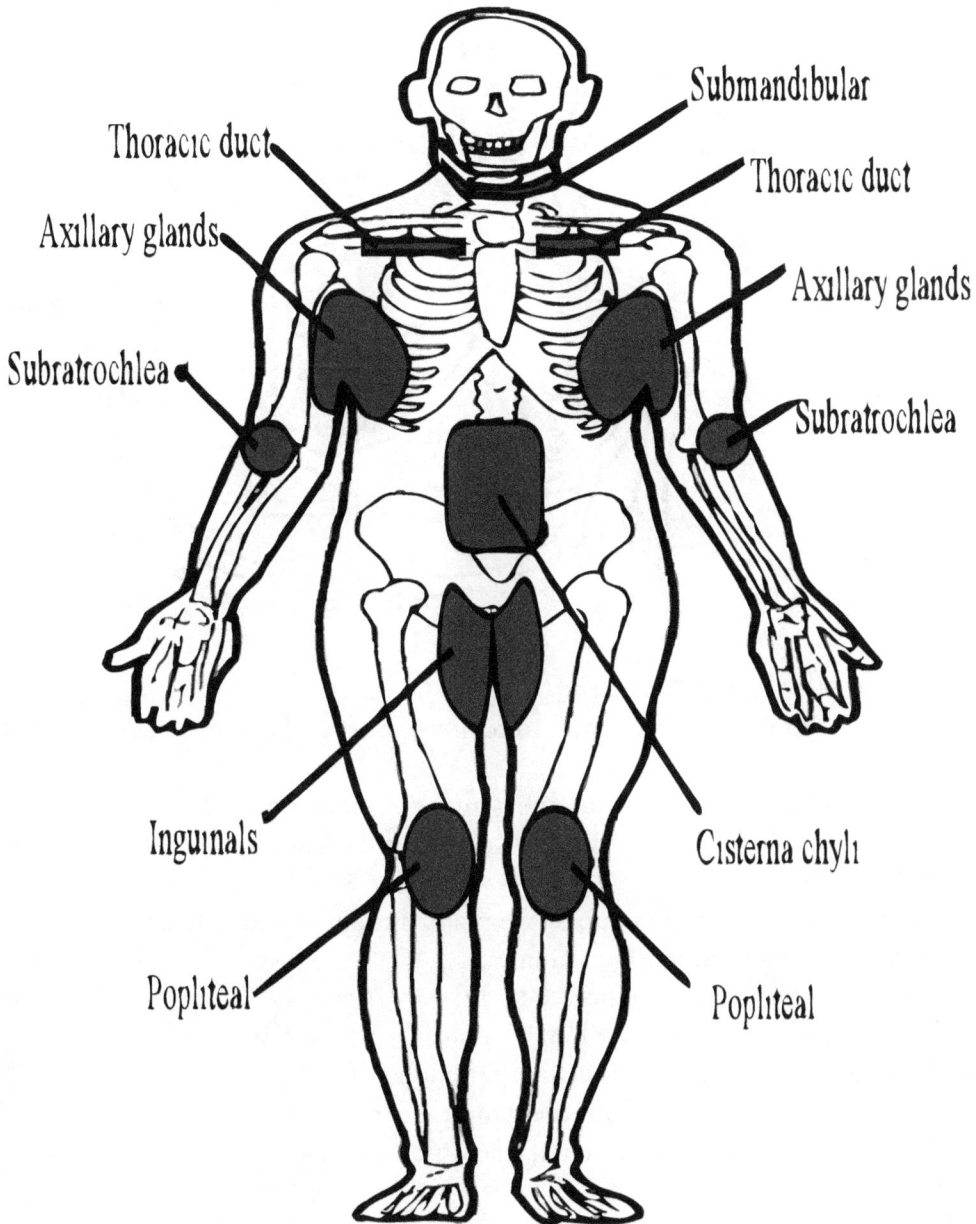

A lymph node consists of several points of entry and one point that exits, the points of entry are called Afferent lymphatic angions and the exiting lymph angion is called Efferent. There is an artery leading to the node and a vein leading out that supplies the lymph node with blood, the point of entry and exit into and out of the lymph node is called the helum. The node is split-up into areas called sinuses that are separated by walls called trabecula.

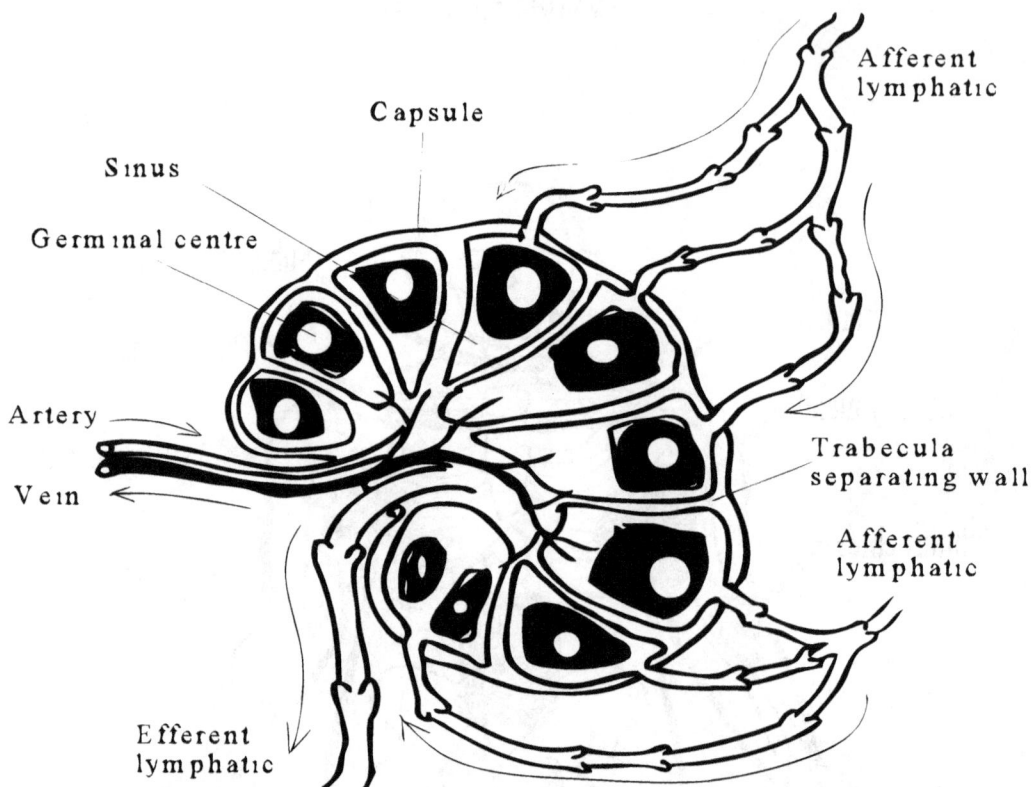

Afferent lymphatic

Capsule

Sinus

Germinal centre

Artery

Vein

Trabecula separating wall

Afferent lymphatic

Efferent lymphatic

The Function of Lymph:
1. To filter out dust particles, and to remove, dyes, pathogens, waste and dead or dying cells.
2. The lymph nodes concentrate the lymph by removing 40 % of the fluid.
3. A lymph node can repair itself provided that there is a blood and nerve supply still intact.
4. To assist in immunity as one of the major sites of immune reactions..
5. Helps maintain blood fluid volume.
6. To circulate proteins around the body.
7. To transport fats, especially long chain cells and proteins.
8. To maintain connective tissue in a functional healthy state.
9. To remove excess water from the lymph via the veins.
10. To store or trap inorganic matter like dust particles.

Problems That Cause Restriction of Lymphatic Flow:
1. Chronic dehydration.
2. Elasticated clothing.
3. Cold will decrease lymph flow.
4. Heat above 104 degrees.
5. Night time or prolonged rest periods will shut the lymph system down.
6. Inflammation will close the lymph valves.
7. Paralysis.
8. Nerve lesions

The Effects of Lymphoedema:
1. Lowered lymph transport capacity.
2. Unable to break down proteins.
3. Swelling in subcutaneous tissue.
4. Excess protein build up.
5. Swelling gets worse as time goes on.
6. In time proteins break down.
7. Fibrosis is the final result.

What Is Lymphoedema:
In long standing oedema the vascular action is increased and the body will lay down more connective tissue and fat cells. Gradually, the tissue becomes hard and lymph action ceases. The appearance is normal, but the skin is taut. Pressing a finger causes a depression that remains after the finger is removed.

The Four Major Components of Lymph:
1. Water   2. Proteins   3. Fats   4. Cells

Three Minor Components of Lymph:
1. Dust   2. Dyes   3. Pathogens

Different Grades of Oedema:
Grade 1: Is reversible, will recover with bed rest and appears soft and will pit with pressure

Grade 2: Is irreversible, will not recover with bed rest and stronger pressure is required to cause pitting, fibrosis occurs, collagen and body connective tissue hardens.

Grade 3: Severe and advanced oedema, sclerosis, elephantiasis, fibrosis around vessels and nerves causing damage, quite often lymphangiosarcoma forms.

LYMPHOSTATIC OEDEMA, is a high protein oedema that can be organic in origin, which means that there occurs a change in the pathway of the lymph at a body structure level. This change in the pathway often means that this type of oedema is often irreversible and cannot be cured. This type of oedema is called **ORGANIC OEDEMA**.

The other type of high protein lymphostatic oedema is called **FUNCTIONAL OEDEMA**. This type of oedema is treatable, and is caused by changes in the function of the body. Some of these changes are due to smooth muscle impairment, skeletal muscle paralysis, angion valve impairment or ruptured collagen filaments.

PRIMARY LYMPHOEDEMA is a branch of organic lymphoedema which normally manifests itself on one limb even though the problem exists in both limbs. It effects the lower extremities more than the upper limbs, and begins distally with a proximal spread. Lymph collector deficiency is the main cause, either congenital or hereditary. There are two types of Primary Lymphoedema, Precox and Tarda. Precox is more commonly found at an early age during puberty and adolescents. Tarda effects the remaining 17% after the age of 30 years, some of the effects are pale thin skin leading to hypokeratosis.

SECONDARY LYMPHOEDEMA is also a branch of organic lymphoedema. This type of oedema is caused by outside intervention like; Parasite (nematode worm) contracted by tropical mosquitoes that infect the blood and in time worms develop and live in the lymph giving rise to a disorder called filariasis disease. Surgery is also another major contributor of secondary oedema, where lymph nodes are removed or damaged. Cancers like, lymphangiosarcoma or Stewart-Treves Syndrome, or tumours, and lastly, constricting clothing like tight socks with elasticated tops, bras with wire or bone supports, underpants with elasticated or tight fitting legs seams. Any of these garments can cause inflammation in the area of constriction.

DYNAMIC OEDEMA unlike lymphostatic oedema is not a high protein oedema, but is in fact a low protein oedema that cannot be managed by manual lymphatic drainage treatment because it is caused by congested heart failure, varicose veins, kidney failure, right side heart failure (this causes blood in the veins to pool causing Veins congestion).

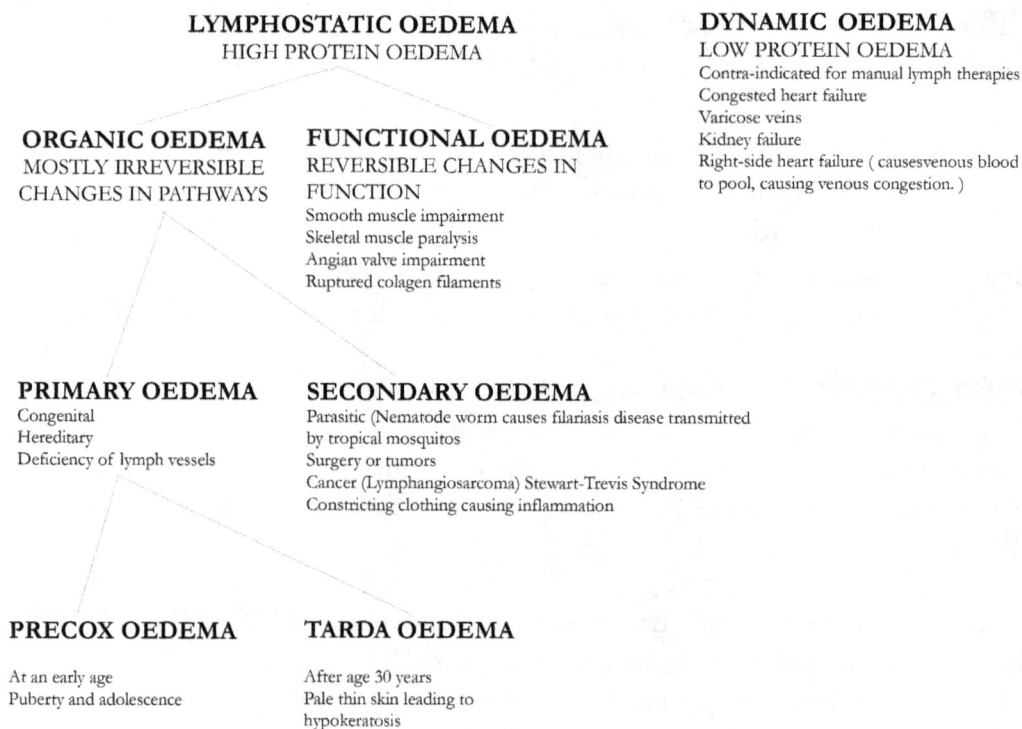

LYMPHOSTATIC OEDEMA
HIGH PROTEIN OEDEMA

DYNAMIC OEDEMA
LOW PROTEIN OEDEMA
Contra-indicated for manual lymph therapies
Congested heart failure
Varicose veins
Kidney failure
Right-side heart failure (causesvenous blood to pool, causing venous congestion.)

ORGANIC OEDEMA
MOSTLY IRREVERSIBLE
CHANGES IN PATHWAYS

FUNCTIONAL OEDEMA
REVERSIBLE CHANGES IN
FUNCTION
Smooth muscle impairment
Skeletal muscle paralysis
Angian valve impairment
Ruptured colagen filaments

PRIMARY OEDEMA
Congenital
Hereditary
Deficiency of lymph vessels

SECONDARY OEDEMA
Parasitic (Nematode worm causes filariasis disease transmitted by tropical mosquitos
Surgery or tumors
Cancer (Lymphangiosarcoma) Stewart-Trevis Syndrome
Constricting clothing causing inflammation

PRECOX OEDEMA
At an early age
Puberty and adolescence

TARDA OEDEMA
After age 30 years
Pale thin skin leading to
hypokeratosis

Although the lymphatic system appears primitive when compared to the other systems, it is in fact an amazing structure that works remarkably well. Every cell in the body must be supplied with nutrition, oxygen, hormones, water, essential minerals and protein. The metabolism requires not only the delivery of these substances but also the efficient removal of waste and other by-products, that's were the lymphatic system comes in.

The lymphatic system starts with tiny subcutaneous pre-lymphatic lymph vessels that cover the sub-surface of the entire body in zones and territories. These areas are defined and mapped out using watershed lines to show us the cut-off point for lymph travelling in a particular direction as seen below.

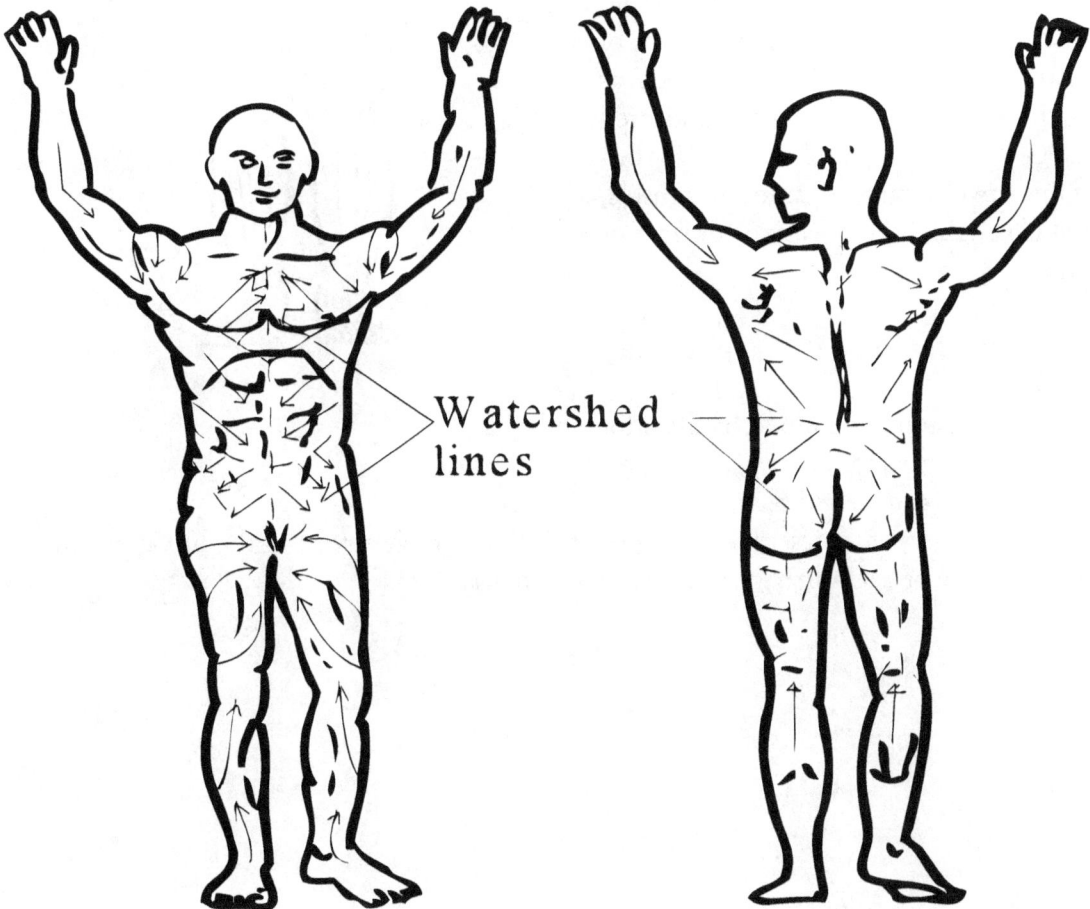

Watershed lines

The Spleen

The spleen is a mass of lymphoid tissue located in the upper left hypochondriac region of the abdomen and is normally protected by the lower part of the rib cage. It sits high under the dome of the diaphragm. The white pulp produces lymphocytes and is scattered throughout the red pulp, in which macrophages filter out old or damaged cells, debris and bacteria. The spleen is also a store of iron from the broken red blood cells and like other lymphoid tissues, is an aid in the development of immunity. The capsule of the spleen, as well as its structure is more elasticated than that of the lymph nodes. It contains involuntary or smooth muscle, which enables the spleen to contract or swell.

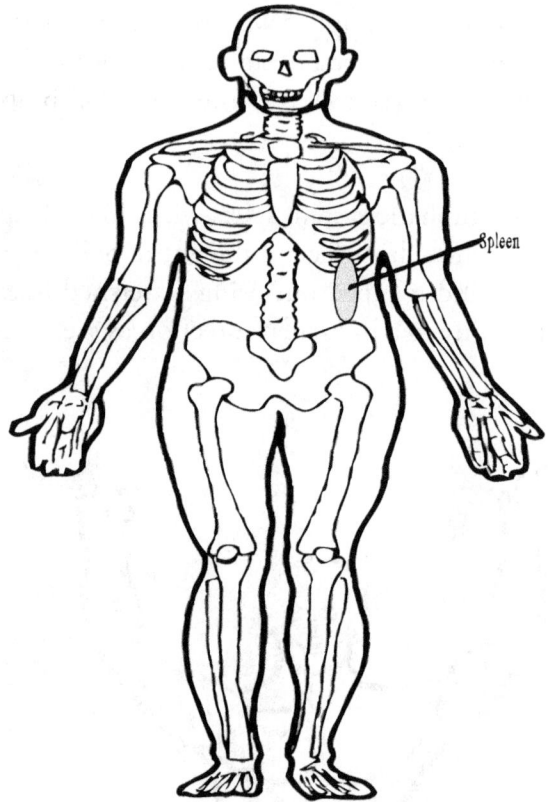

The major functions of the spleen are as follows:
1. Cleaning the blood by filtration and phagocytosis.
2. Destroying old or worn-out red blood cells.
3. To produce red blood cells before birth.
4. To serve as a reservoir for blood, which can be returned to the bloodstream in case of haemorrhage or other emergencies.

Spleen

Some Conditions of The Lymphatic System

Disease or Disorder   **ELEPHANTIASIS**

Description   A great enlargement of an extremity resulting from blockage of lymphatic pathways. Usually this is the end stage of a lesion of filariasis, that has possibly lasted for years, it is characterized by tremendous swelling, usually of the external genitalia and legs. The overlying skin becomes dark, thick, and coarse.

Possible Allopathic Treatment   Not available at present. An option would be surgery and diuretics.

Disease or Disorder   **LYMPHADENITIS (lim-fad-en-i-tis)**

Description   Inflammation of the lymph nodes, the nodes become enlarged and tender. Lymph node involvement may be generalized, with systemic infections or remain local. This condition reflects the body's attempt to combat an infection. Cervical lymphadenitis occurs during measles, scarlet fever, septic sore throat, diphtheria and frequently, the common cold. Chronic lymphadenitis may be due to the tubercle bacillus (TB). Infections of the upper extremities cause enlarged axillary nodes, as does cancer of the mammary glands. Infection of the external genitals or the lower extremities may cause enlargement of the inguinal lymph nodes.

Possible Allopathic Treatment   Treatment depends on underlying cause, hot, wet applications may help relieve pain. A lymph node biopsy will be necessary to determine the treatment.

Disease or Disorder   **LYMPHADENOPATHY (limph-aden-opa-thee)**

Description   Lymphadenopathy is a term meaning "disease of the lymph nodes", Enlarged lymph nodes are common symptom in a number of infectious and cancerous diseases. An early sign of infection with HIV, the virus that causes AIDS, is a generalized lymphadenopathy. Infectious mononucleosis is an acute viral infection the hall mark of which is a marked enlargement of the cervical lymph nodes. Mononucleosis is fairly common among college students.

Possible Allopathic Treatment   Allopathic treatment would be too varied to specify here due to the onset of the disease, its progression and its cause. Treatment depends on underlying cause, but a hot, wet applications may help relieve pain. A lymph node biopsy will be necessary to determine the treatment.

Disease or Disorder	**SPLENOMEGALY**
Description	An abnormal enlargement of the spleen, accompanies certain acute infectious diseases, including scarlet fever, typhus fever, typhoid fever and syphilis. Many tropical parasitic diseases cause splenomegaly. A certain blood fluke (flatworm) that is fairly common among workers in Japan and other parts of Asia causing marked splenic enlargement.
Possible Allopathic Treatment	Allopathic treatment would be too varied to specify here due to the onset of the disease, its progression and its cause. Treatment depends on underlying cause.

Disease or Disorder	**SPLENIC ANAEMIA**
Description	Characterized by enlargement of the spleen, haemorrhages from the stomach, and accumulation of fluid in the abdomen.
Possible Allopathic Treatment	Allopathic treatment would be too varied to specify here due to the onset of the disease, its progression and its cause. Treatment depends on underlying cause, But in this disease and others of the same nature, splenectomy appears to constitute a cure.

Disease or Disorder	**HODGKIN'S DISEASE**
Description	Hodgkin's Disease is a chronic malignant disorder, most common in young men, characterized by painless enlarged lymph nodes. The nodes in the neck particularly, and often those in the armpit, thorax and groin. The spleen may become enlarged as well. The present of Reed-Sternberg cells, this is a large, abnormal multinucleated reticuloendothelial cell found in the lymphatic system. Symptoms include anorexic type weight loss, generalized pruritis, low-grade fever, night sweats, anaemia and leucocytosis.
Possible Allopathic Treatment	Chemotherapy and Radiation either separately or in combination, have been used with good results 50% of the time, affording patients many years of life.

Disease or Disorder	**LYMPHOSARCOMA**
Description	Lymphosarcoma is a malignant tumour of the lymphoid tissue that is likely to be rapidly fatal. Fortunately, it is not a common disease. (Lymphoma, is any tumour, benign or malignant, that occurs in lymphoid tissue). Effects similar to Hodgkin's Disease.
Possible Allopathic Treatment	Early surgery in combination with appropriate radiotherapy offers the only possible cure at this time.

The Neurological System

Introduction

The neurological system is composed of nerve cells and supporting tissue. This control system has the ability to unify the body through various functions to maintain homeostasis. These functions are controlled by sending, receiving and carrying, nerve impulses. The whole neurological system is based on chemical conductive materials related directly to the nerves and axons. The nerve cells are very sensitive and their fibres specialize in the transmission of impulses, this network runs throughout the body with a two-way connection.

The neurological system serves three main functions:
1. **Sensory functions** - it will sense changes both within the body's internal and external environment.
2. **Integrative functions** - the body's ability to interpret these sensed changes
3. **Motor functions** - the body's ability to respond to the interpreted changes and start the necessary action through either muscular contractions or glandular secretions.

The neurological system has two main divisions:
1. **Central** (cerebrospinal) **nervous system**. (CNS)
2. **Peripheral nervous system**. (PNS)

The Central Nervous System

This is the control centre for the entire body. It consists of the brain and the spinal cord. The brain functions as the command centre while the spinal cord serves as an extension of the brain. The CNS uses both ascending and descending impulses to transfer the information it receives and gives to the PNS. It serves as a source of motor commands for muscles and receives sensory input from the Peripheral Nervous System.

The Peripheral Nervous System

This system is composed of a network of both cranial and spinal nerves (neuron processes) and ganglia (islands of neurons) that connect the CNS to other areas of the body. They are located outside of the CNS and include the Afferent (Sensory) System, the Efferent (Motor) System, the Somatic Nervous System, and the Autonomic Nervous System (which includes the Sympathetic and Parasympathetic systems.

Nerves of the PNS generally have both fibres from the somatic system, which generates voluntary response and the autonomic system, which controls involuntary response. There are twelve pairs of cranial nerves and thirty one pairs of spinal nerves. These branches of spinal nerves are referred to as peripheral nerves. Their function is to relay sensations from the body to the brain and spinal cord and relay motor commands to all skeletal muscles.

CENTRAL NERVOUS SYSTEM

|

BRAIN & SPINAL CORD

|

CONTROLS MOTOR COMMANDS AND SENSORY INPUT

PERIPHERAL NERVOUS SYSTEM

|

CONTROLS AUTONOMIC NERVOUS FUNCTIONS

|

INCLUDES SYMPATHETIC & PARASYMPATHETIC FUNCTIONS

How Is this Info Relayed?

The basis for the nervous system is the nerve cell or neuron. These pass the information to the CNS or PNS. They consists of a nerve cell body that has a receiving mechanism (dendrite) and a transmitting mechanism (axon).

Neurons are classified into three functional types. These include:

1. **Sensory or Afferent neurons** - These are responsible for conducting impulses from receptors in the body to the CNS. They pick up sensations such as touch, pressure, pain, joint position, muscle tension, etc.

2. **Motor or Efferent neurons** - These direct command impulses from the CNS to the muscles via the motor end plate (a band of fibres that flow into the sheath of the muscle) and releases a neurotransmitter to induce muscle action.

3. **Inter-neurons** - These are found within the CNS. They carry information throughout the CNS and help transmit impulses from sensory neurons to motor neurons.

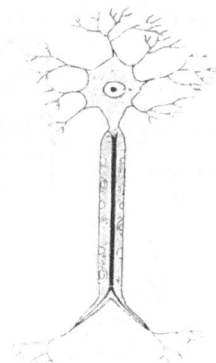

There are two types of peripheral nerve fibres.
- White nerve fibres which are myelinated (enclosed in a sheath) .
- Grey nerve fibres which are un-myelinated (without a sheath).

Myelinated nerve axon

The function of the myelin sheath is to increase the speed of the nerve impulse and to insulate and maintain the nerve fibre (axon).

A nerve impulse is required to conduct the information from one neuron to another until it reaches the required destination. Neurons do not have contact with one another, there is a space between them. This gap must be crossed. It is done through a "bridging of the synapsis gap". This involves the nerve endings to liberate chemical substances which stimulate the adjacent cell to start a fresh impulse along its own fibre. This process continues throughout the nerve network.

Example:
A hand is placed on a hot stove. The sensory neurons and the pain receptors of the PNS register the excessive heat and conducts a message through the neurons to the CNS and the brain. The brain processes the information and sends a message back down the spine, out of the CNS, through the motor neurons of the PNS, which in turn will transmit the information to the muscles of the hand in order to contract the muscles to pull away from the heat.

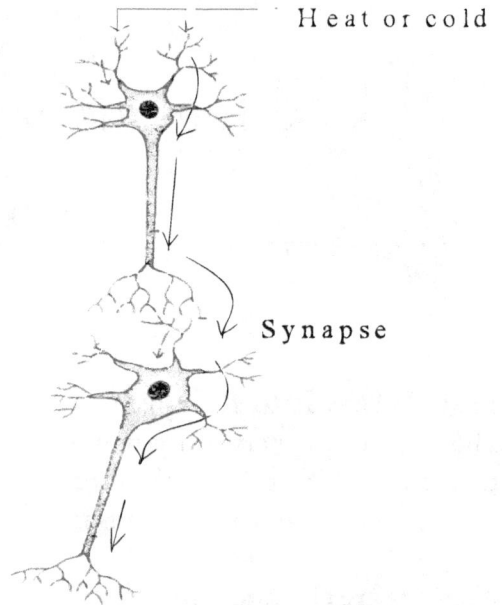

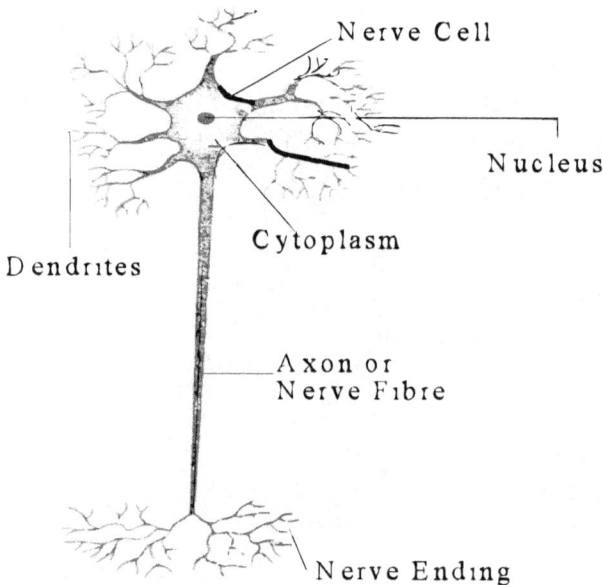

Heat or cold

Synapse

Nerve Cell

Nucleus

Cytoplasm

Dendrites

Axon or Nerve Fibre

Nerve Ending

The chemical mediators used to create synapsis include acetylcholine and adrenaline (epinephrine and norepinephrine).

Acetylcholine is a skeletal muscle chemical mediator. This neurotransmitter is widely distributed in the body tissues. After it has been utilized, it is then destroyed by an enzyme called acetylcholinesterase. Its function is to inactivate the acetylcholine released during nerve impulses and further prevent accumulation by breaking down the acetylcholine into acetate and choline.

Both acetylcholine and adrenaline are an autonomic neurotransmitters. When we consider the autonomic system, we shall see that chemical agencies play a very large part in regulating internal organs and blood vessels and that certain endocrine glands, like the adrenal, can secrete the same substances into the blood as hormones to secure a rapid response in emergencies.

Autonomic Nervous System

The autonomic nervous system functions without conscious control. Its role is to maintain homeostasis within the body by manipulating the activity of the visceral organs. The autonomic nervous system will either increase or decrease activity according to changing internal conditions. Some of the functions it is responsible for include regulation of the heart rate, body temperature, blood pressure, etc. It is divided into two separate parts: the sympathetic system and the parasympathetic nervous system.

TABLE OF AUTONOMIC FUNCTIONS

EFFECTOR	SYMPATHETIC CONTROL	PARA SYMPATHETIC CONTROL
Heart muscle	Accelerates heartbeat	Slows heart beat
Smooth muscle of most blood vessels	Constricts blood vessels	None
Smooth muscle of blood vessels in skeletal muscles	Dilates blood vessels	None
Smooth muscle of the digestive tract	Decreases peristalsis & inhibits defecation	Increases peristalsis
Smooth muscle of the anal sphincter	Stimulates - closes sphincter	Inhibits - opens sphincter for defecation
Smooth muscle of the urinary bladder	Inhibits - relaxes bladder	Stimulates - contracts bladder
Smooth muscle of the urinary sphincter	Stimulates - closes sphincter	Inhibits - opens sphincter for urination
Smooth muscle of the iris	Stimulates radial fibres - dilation of pupil	Stimulates circular fibres - constriction of pupil
Smooth muscle of the ciliary	Inhibits - accommodation for far vision (flattening of lens)	Stimulates - accommodation for near vision (bulging lens)
Smooth muscle of hairs (pilomotor muscles)	Stimulates - goose pimples	No parasympathetic fibres
Adrenal medulla gland	Increases epinephrine secretion	None
Sweat glands	Increases sweat secretion	None
Digestive glands	Decreases secretion of digestive juices	Increases secretion of digestive juices

All the internal organs have a double nerve supply from the sympathetic and parasympathetic systems and their effect is opposite - simply, a sympathetic nerve has the effect of increasing body activity and speeds it up, whereas the parasympathetic, on the contrary, slows down body activity. The sympathetic system comprises a ganglionic cord which runs anteriorly on either side of the vertebral column. The principle plexus of this system is the cardiac plexus. This supplies all the thoracic viscera and the thoracic vessels. The coeliac or solar plexus supplies all the abdominal viscera and the hypogastric plexus supplies the pelvic organs. The effects of sympathetic stimulation is increased body activity in relation to fear, flight or fight, aided by adrenaline secreted by the adrenal gland.

Sympathetic stimulation

1. Constricts the blood vessels of the skin, which raises the blood pressure
2. Shunts blood to the heart and brain,
3. Speeds and strengthens the heartbeat,
4. Dries up glandular secretions,
5. Dilates the pupils,
6. Stands the hair on end,
7. Initiates sweating,
8. Mobilizes glucose, and relaxes the walls of the hollow viscera.

The sympathetic system provides for today's work and that its action increases when involved with physical activity.

The parasympathetic nervous system consists mainly of the vagus nerve which gives off branches to the organs of the thorax and abdomen, but also includes branches from other cranial nerves, mainly the third, seventh, and ninth as well as nerves in the sacral region of the spinal column. The parasympathetic system produces relaxed states, it dilates the peripheral vessels, slows the heart and lowers the blood pressure and excites secretion and peristalsis. The parasympathetic system looks after tomorrow, being mainly concerned with the changes that takes place during rest.

In some people, the sympathetic nerves are stronger and hold the balance in the body. The sympathetic nerves are stimulated by strong emotions such as fear, anger and excitement. The adrenal is one of the glands which the sympathetic system stimulates and the liberation of adrenalin is one of the body's responses to strong emotions.

In other people, the parasympathetic nerves are stronger and hold the balance in the body. These people have a placid disposition, good digestion and are not very easily disturbed. These are known as vagotonic types. In other people, the sympathetic nerves are the stronger and these people are more emotional, less stable and their digestion is more readily disturbed. These are known as sympatheticotonic types.

Within the autonomic nervous system, there are peripheral ganglia. These ganglion connect a paired unit of autonomic motor systems. One neuron from the CNS and one from the PNS. These are divided into pre-ganglionic fibres (they enter the ganglion) and post ganglionic (they exit the ganglion).

Pre-ganglionic fibres of the sympathetic and parasympathetic systems secrete acetylcholine (epinephrine), but sympathetic postganglionic chemical neurotransmitters secrete norepinephrine (adrenaline) while the parasympathetic secrete acetylcholine.

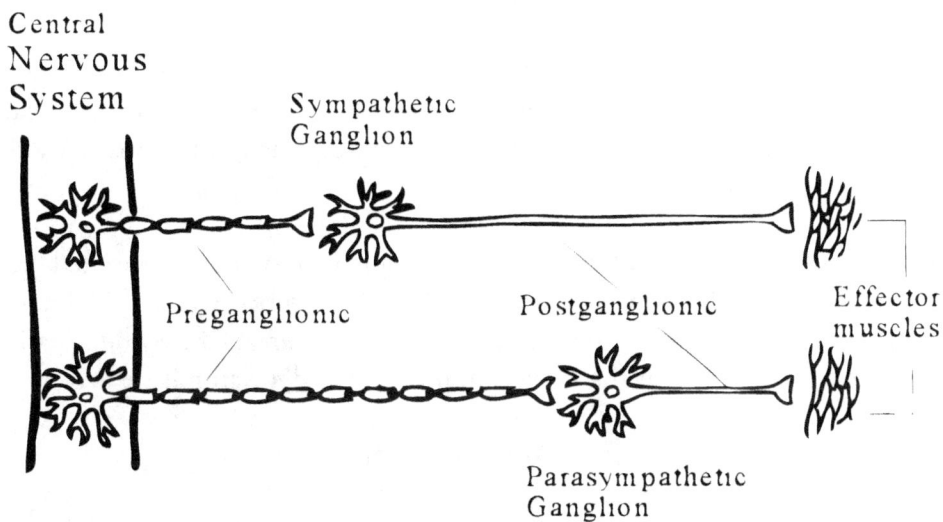

Central Nervous System — Sympathetic Ganglion — Preganglionic — Postganglionic — Effector muscles — Parasympathetic Ganglion

Central Nervous System

Brain

Afferent (sensory) System
Conveys information from receptors to the central nervous system

Somatic Nervous System
Conveys information from the central nervous system to skeletal muscles

Spinal Cord

Efferent (motor) System
Conveys information from the central nervous system to muscles and glands

Autonomic Nervous System
Conveys information from the central nervous system to smooth muscle, cardiac muscles and glands

Sympathetic Nervous System

Stimulates or increases organ activity constricts blood vessels, raises blood pressure, dilates pupils, etc.

Parasympathetic Nervous System

Inhibits or decreases activity Dilates vessels, slows heart, lowers blood pressure, etc.

Sympatheticotonic type

Vagotonic type person

Dermatomes

Dermatomes are areas of the skin that are supplied with sensory receptors that are connected and enervated by a specific spinal nerve. These dermatomes only communicate cutaneous pain and any pain that is referred to the skin.

If the nerve supply is interrupted, there may be a loss of sensation in that area. However, many dermatomes overlap in certain areas (especially in the trunk), therefore there may be no loss of sensation.

Many professional practitioners know which spinal nerves are associated with each dermatome. With this knowledge, it is possible to determine which spinal nerve is not functioning appropriately by stimulating a dermatome area. If the client/patient feels no sensation, it may be that the nerves supplying that dermatome are involved.

DERMATOMES

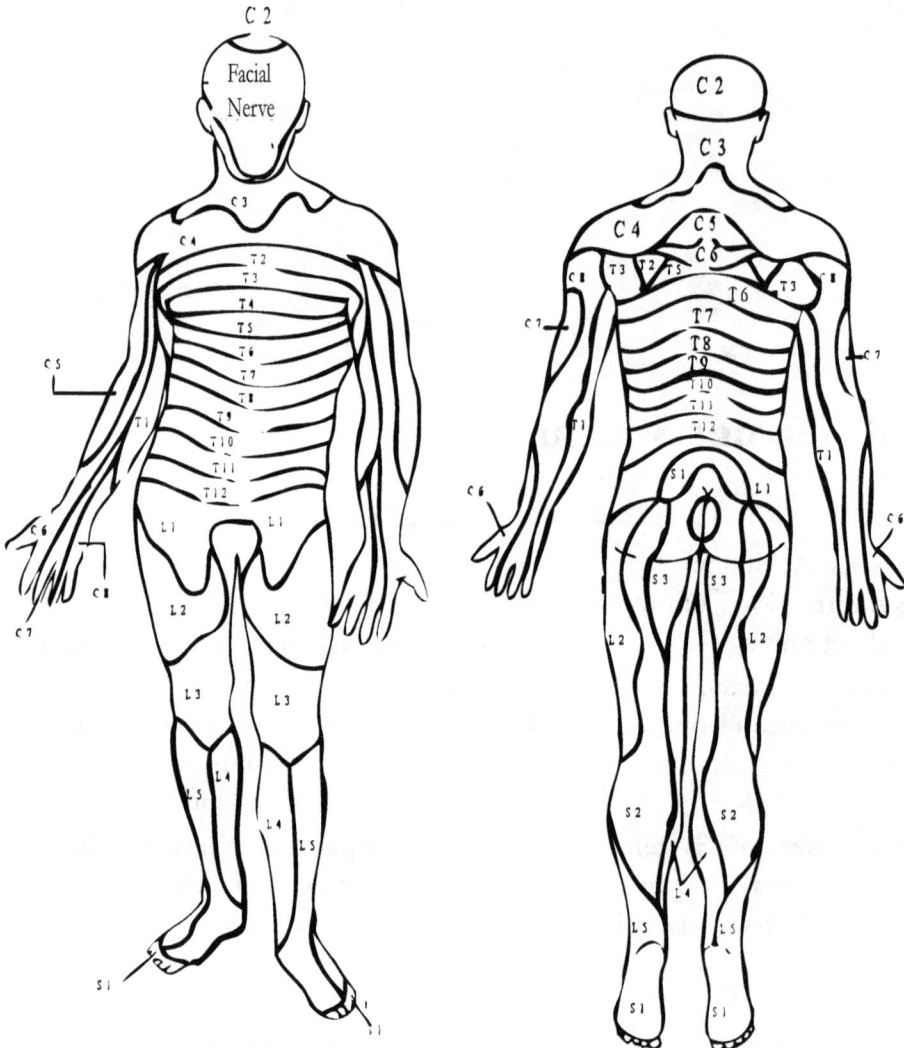

The Brain

At the centre of the nervous system is the brain. The brain is well protected from the outside by the hard bone structure of the skull. Inside the brain is protected externally by three enveloping membranes known as the **meninges**.

The **outer layer** is known as the **dura-mater** or strong or hard mother. A tough loosely applied protective envelope, in the cranium, it also forms the lining periosteum of the skull.

The **middle layer** is known as the **arachnoid**. This fits closely inside the dura, but there is a sub-arachnoid space separating it from the pia, filled with cerebrospinal fluid and traversed by spidery connective tissue.

The **inner layer** is called the **pia mater** or **soft mother**, a fine membrane closely applied to the brain and cord, following every cleft and crevice and carrying with it the fine blood vessels.

The outer meninges (the dura mater) is constructed of strong fibrous tissue anchored to the skull. The middle tissue, or arachnoid, is much more delicate and is not anchored to the skull, thus allowing the brain to expand. Under it lies the big reservoir of cerebral spinal fluid by which the whole of the brain is surrounded and on which it rests. Then comes the pia mater which is in contact with the grey matter of the brain itself and dips deep down between the brain convolutions.

When we speak of the brain, we are really considering three quite different structures.

THE CEREBRUM, THE CEREBELLUM AND THE MEDULLA OBLONGATA.

The adult human brain weighs rather more than three pounds and is so full of water that it tends to slump rather like a jelly if placed without the support of a firm surface. It is estimated that it has twelve billion neurons or nerve cells.

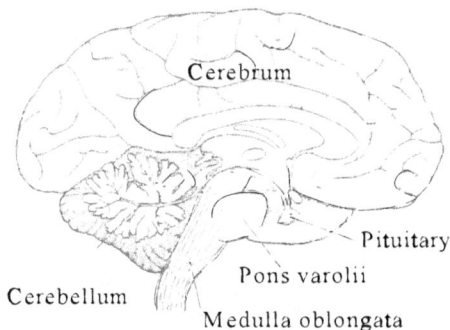

Cerebrum

Pituitary

Pons varolii

Cerebellum

Medulla oblongata

The Cerebrum

The cerebrum consists of two symmetrical hemispheres. The outer layer of the cerebrum is known as the cortex and this is arranged in convolutions, that is, deep irregular shaped fissures or indentations. This is the grey matter of the brain. Underneath the cortex lies nerve fibre or white matter.

The function of the cerebrum is to control voluntary movement, to receive and interpret conscious sensations and it is the seat of the higher functions such as the senses, memory, reasoning, intelligence, and moral sense.

The Cerebellum

The cerebellum is much smaller in size and lies below and behind the cerebrum. It too has grey matter under which is white matter. Its function is to control muscular co-ordination and balance.

The Medulla Oblongata

The medulla oblongata is about 32mm long, tapering from its greatest width of 9mm and connects the rest of the brain with the spinal cord with which it is continuous. It is made up of interspersed white and grey matter. The medulla oblongata not only acts as the link between the brain and the central nervous system of the body, but it is also the centre of those parts of the autonomic nervous system which controls the heart, lungs, processes of the digestion, etc.

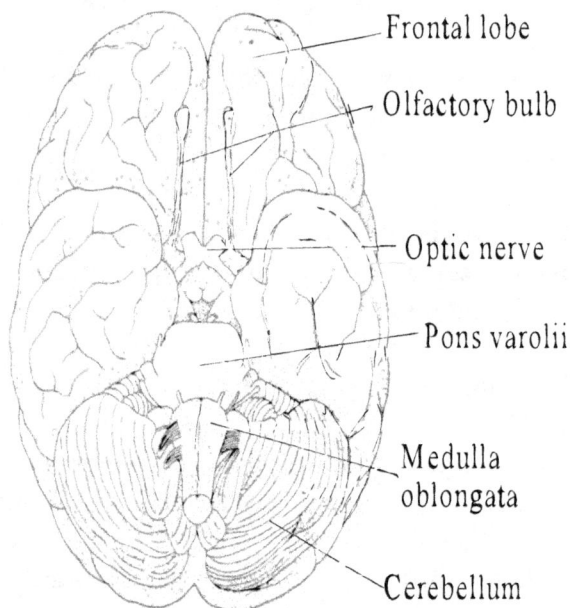

Frontal lobe

Olfactory bulb

Optic nerve

Pons varolii

Medulla oblongata

Cerebellum

The Spinal Cord

The spinal cord, which is continuous with the medulla oblongata, extends downwards through the vertebrae of the spinal column. The cord itself is cylindrical in shape with an outer covering of supporting cells and blood vessels and an egg-shaped core of nerve fibres. It has two swellings, the cervical and lumbar enlargements, which are the origins of the roots of the brachial and lumbar plexuses for the upper and lower limbs. It extends through four-fifths of the spinal column and averages between sixteen and seventeen inches in length.

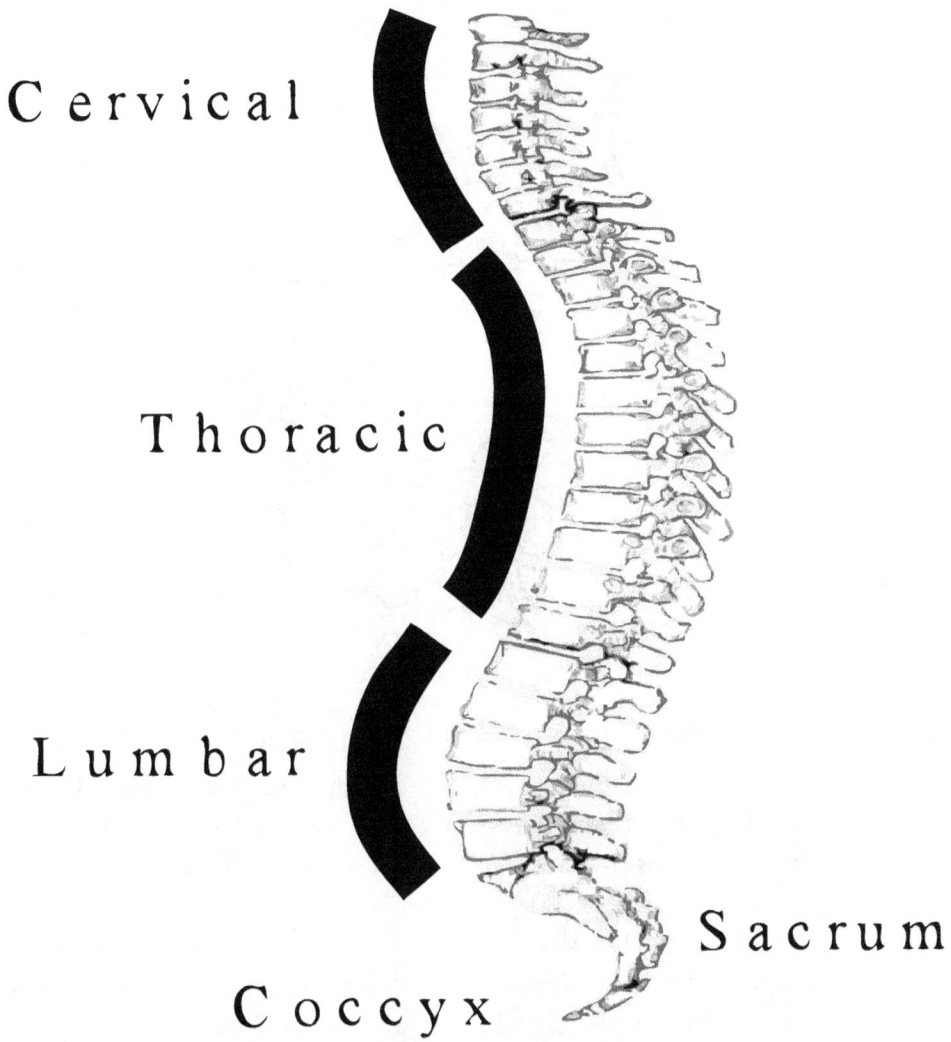

Cervical

Thoracic

Lumbar

Sacrum

Coccyx

STRUCTURE GOVERNS FUNCTION:

C=CERVICAL   T=THORACIC   L=LUMBAR

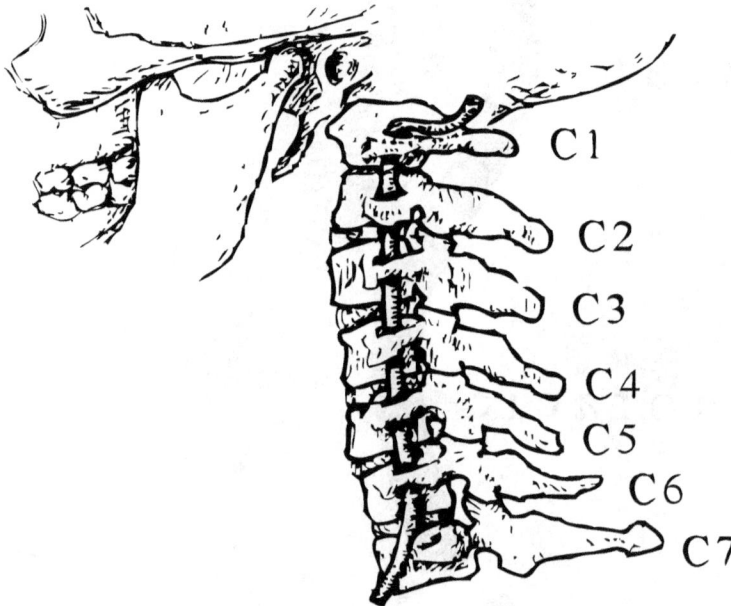

VERTEBRAE	AREA AFFECTED	SYMPTOMS
C1	Blood supply to the head Pituitary Gland & brain, Sympathetic Nervous System,	Headaches, Nervousness. Insomnia, Mental Conditions. Amnesia, Epilepsy, Tiredness.
	Ear both middle and inner.	Dizziness, St. Vitus Dance
C2	Optic & Auditory Nerve Sinuses, Mouth, Forehead	Allergies, Deafness, Eye trouble, Earache, Fainting Spells
C3	Cheeks, Outer ear, Face Bones, Teeth, Tri-facial Nerve	Neuralgia, Acne, Neuritis Eczema
C4	Nose, Lips, Mouth Hearing, Adenoids Eustachian Tube	Hay Fever, Catarrh, Ear ache, Sore throat
C5	Vocal cords, Neck Glands	Laryngitis, Throat conditions
C6	Neck Muscles, Shoulders Tonsils	Stiff Neck, Upper arm pain, Tonsillitis, Cough & Croup
C7	Thyroid	Bursitis of the shoulder or elbow, Colds, Thyroid Conditions, Goitre.

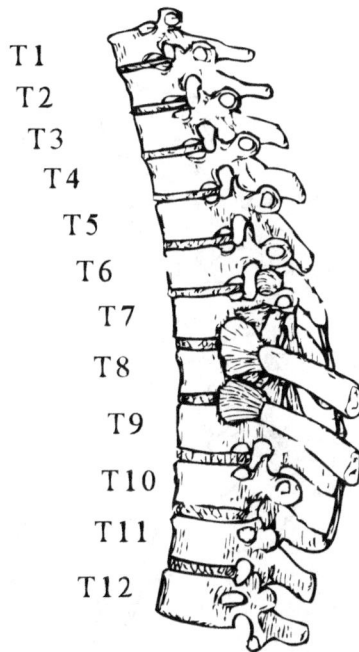

T 1
T 2
T 3
T 4
T 5
T 6
T 7
T 8
T 9
T 10
T 11
T 12

VERTEBRAE	AREA AFFECTED	SYMPTOMS
T1	Lower Arms, Oesophagus Trachea	Asthma, Cough, Difficult Breathing, Pain in lower arm and hands.
T2	Heart & Valves Coronary Arteries	Functional heart conditions, Chest pains
T3	Lungs, Bronchial Tube Pleura, Chest, Breast	Bronchitis, Congestion, Flu, Pleurisy, Grippe, Pneumonia
T4	Gall Bladder, Common Duct	Gall Bladder Conditions Jaundice, Shingles.
T5	Liver, Solar Plexus Blood Pressure	Liver Conditions, Fever, Arthritis, Anaemia, Poor circulation
T6	Stomach	Heartburn, Dyspepsia, Indigestion, Stomach Troubles
T7	Pancreas, Duodenum	Diabetes, Ulcers, Gastritis
T8	Spleen, Diaphragm	Leukaemia, Hiccoughs.
T9	Adrenals	Allergies, Hives
T10	Kidneys	Kidney Troubles, Hardening of the Arteries, Chronic Tiredness, Nephritis, Pyelitis
T11	Kidneys, Ureters	Skin Conditions like acne & Eczema
T12	Small Intestines, Fallopian Tubes,	Rheumatism, Flatulence Lymph Circulation

L 1
L 2
L 3
L 4
L 5

S a c r u m
X 5

C o c c y x
X 4

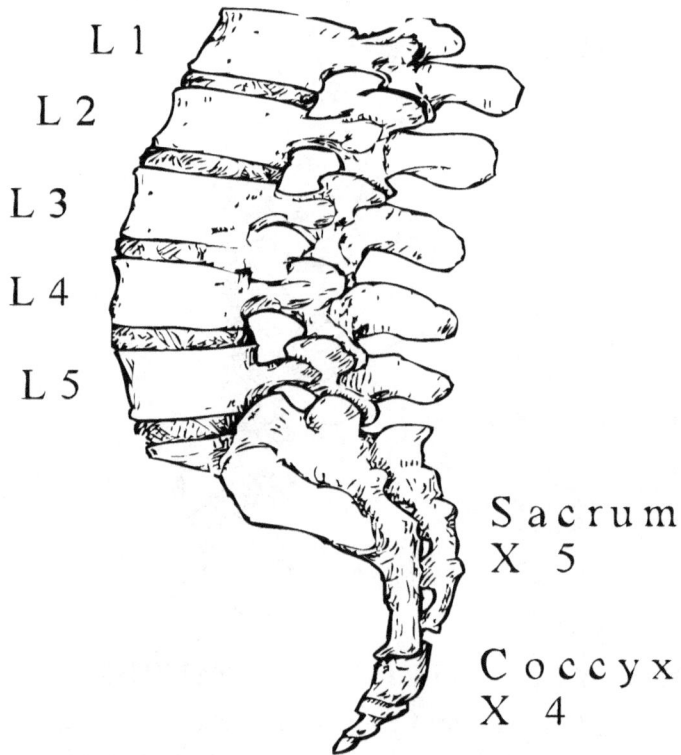

VERTEBRAE	AREA AFFECTED	SYMPTOMS
L1	Large Intestines	Constipation, Colitis, Hernia, Diarrhoea, etc.
L2	Appendix, Abdomen, Upper Leg, Caecum	Appendicitis, Varicose Veins, Cramps, Difficult Breathing
L3	Sex Organs, Uterus, Bladder, Knee	Bladder Troubles, Menstrual Troubles, Miscarriages, Bed Wetting, Impotency, Change of Life Symptoms, Many Knee pains.
L4	Prostrate Gland Muscles of the Lower Back, Sciatic Nerve	Sciatica, Lumbago, Backache
L5	Lower Legs, Ankles, Feet, Toes, Arches, Sacrum, Buttocks, Hip Bones, Spinal Curvatures, Coccyrectum Anus.	Poor Circulation in the legs, Swollen & Weak ankles, Cold feet, Weak legs, Sacro-Iliac conditions, Haemorrhoids, Pruritus or itching, Pain at the end of the spine

There are twelve pairs of cranial nerves given off from the base of the brain. Thirty-one other pairs branch off the spinal cord throughout its lengths. These extend to every part of the body. Nerves that extend upwards through the spinal cord to the brain pass through the medulla oblongata where they cross, this enables the right side of the brain to control the left side of the body. Nerves that run to and from the central nervous system are known as peripheral nerves. These fall into three categories:

1. MOTOR or EFFERENT NERVES; The primary function of these nerves are to control the movement of muscles.
2. SENSORY or AFFERENT NERVES; These carry impulses from the sensory nerve endings to the spinal column and brain.
3. MIXED NERVES; These consist of both motor and sensory fibres. Cranial nerves.

NAME	TYPE	FUNCTION
Abducent	Motor	Supply lateral rectus muscles of the eyes.
Auditory	Sensory	Sense of hearing, maintenance of balance, equilibrium
Facial	Mixed	Sense of taste, Facial expression muscles.
Glossopharyngeal	Mixed	Sensations from tongue, muscles of the pharynx
Hyperglossal	Motor	Supplies tongue muscles
Oculi Motor	Motor	Supply muscles operating the eyes.
Olfactory	Sensory	Sense of smell.
Optic	Sensory	Sense of sight.
Trochlear	Motor	Supply superior oblique muscles of the eyes.
Trigeminal	Mixed	Receiving pain, heat, pressure and stimulating muscles of mastication.
Spinal Accessory	Motor	To Sterno-Cleido Mastoid Trapezius Muscles
Vagus	Mixed	Sensory, Motor, Digestive and Respiratory organs

The **thirty-one pairs of spinal nerves** comprise eight pairs of cervical nerves, twelve pairs of thoracic nerves, five pairs of lumbar nerves, five pairs of sacral nerves and one pair of coccygeal nerves.

DIAGRAM OF NERVES OF THE BODY

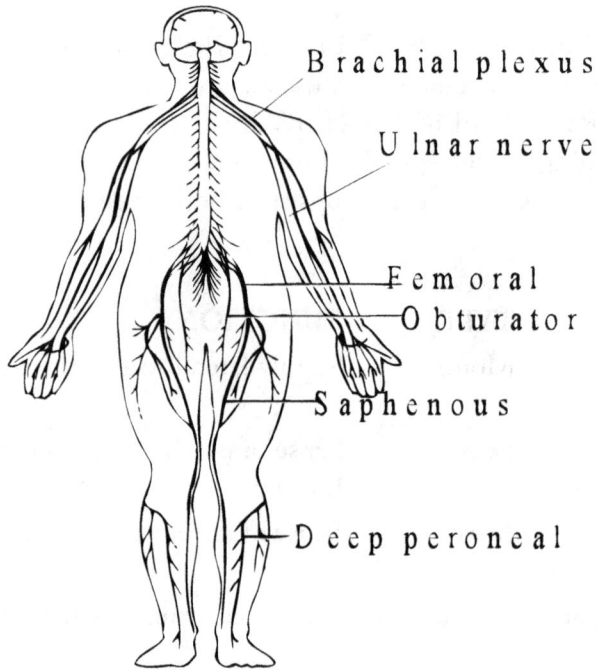

Brachial plexus

Ulnar nerve

Femoral

Obturator

Saphenous

Deep peroneal

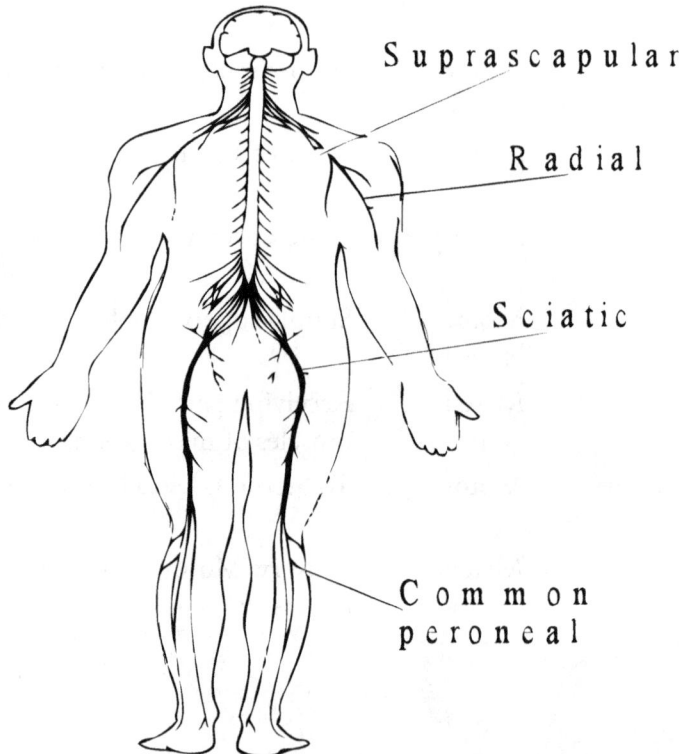

Suprascapular

Radial

Sciatic

Common peroneal

Some Conditions of the Neurological System

Disease or Disorder	**NEURITIS**
Description	Inflammation of a nerve. This takes in a wide group of disturbances which effect the peripheral nerves after they leave the spinal cord. Symptoms include, neuralgia, hyperthesia, anaesthesia, paralysis, muscular atrophy and poor reflexes. Some of the disturbances are due to infection, others to compression of the nerves and probably the biggest single factor is the build up of urea and lactic acid at a point or points of the nerve's course which effect the nerve's sheathing.
Possible Allopathic Treatment	Rest and recuperation, anti-inflammatory drugs and symptomatic treatment in response to prevailing manifestations present.
Disease or Disorder	**BELL'S PALSY or FACIAL PARALYSIS**
Description	Bell's Palsy is named after the Scottish surgeon Charles Bell in 1774. Bell' Palsy is an inflammation of the facial nerve, causing neuritis type symptoms. Often caused by infection, or compression of the nerve as it passes through a tiny opening in the skull inferior and anterior to the ear. Often the results of trauma or nerve compression by a tumour causing partial paralysis on one side of the face. The patient is unable to close an eye or control salivation on one side of the face, the condition is usually unilateral and can be transient or permanent. The extent of the nerve damage will determine the recovery outcome. Complete recovery within 2 months is not uncommon with mild or partial paralysis. The recovery of total paralysis varies from case to case, provided the nerve retains its excitability to maximise electrical stimulation then 90% of those afflicted will eventually recover. If electrical excitability is absent or low, misdirected regrowth of nerve fibres cause unexpected muscle contractions and chronic weakness.
Possible Allopathic Treatment	Methylcellulose drops and eye patches. Prednisone has been used to help relieve acute pain and reduce residual paralysis on occasion. Faradic stimulation of the nerve and physical therapy are useful in provoking muscle motion and preventing muscle contraction. Hypoglossal-facial nerve anastomosis may partially restore facial function between 6-12 months.

Disease or Disorder	**NEURALGIA**
Description	A painful condition in a nerve due to irritation, inflammation, trauma or exposure. Characterized by severe stabbing pains. One of the most common forms of neuralgia, effects the trigeminal nerve, causing facial pain. Glossopharyngeal neuralgia will cause severe pain in the throat to middle-ear area.
Possible Allopathic Treatment	Drugs like, Carbamazepine is generally effective or amitriptyline at bedtime occasionally helps. Surgery may be used to separate structures like arteries from compressing nerves.

Disease or Disorder	**PARKINSONS DISEASE**
Description	Otherwise known as paralysis agitans, this is an extremely common illness beginning in the middle life, deriving from disease of the basal ganglia. The disease is slowly progressive, but it does not effect the brain proper so there is no loss of speech and intelligence is unaffected. The 4 chief symptoms of this illness are resting tremors, muscular rigidity, postural instability and slowness and poverty of movement. Secondary Parkinson results from loss or interference of dopamine within the basal ganglia. The most common cause of S/P is ingestion of neuroleptic drugs that block dopamine.
Possible Allopathic Treatment	Drug therapy like Levodopa, this is a metabolic precursor to dopamine and crosses the blood-brain barrier into the basal ganglia where it is decarboxylated to form dopamine, replacing the missing neural-transmitter. Tricyclic antidepressants are often used to deal with the secondary effects of depression.

Disease or Disorder	**SCIATICA**
Description	Is an inflammation of the great sciatic nerve which is the longest single nerve in the body. This is often a form of rheumatic neuritis but can also be caused by compression by an arthritic spur or a slipped disc, marked by pain & tenderness along the length of the nerve. Recovery from a single acute attack is common, but attacks that recur often become chronic. Acute low back pain following strain or over use of muscles are characterized by muscle tightening, tenderness or spasm.
Possible Allopathic Treatment	In the first instance bed rest to relieve muscle spasm, localised heat pads, massage, oral analgesics, NSAIDs, muscle relaxants. Chronic pain treatment may involve wearing lumbosacral corsets, muscle strengthening exercises and weight loss diets.

Disease or Disorder	**POLIOMYELITIS**
Description	Acute virus infection that destroys the anterior columns of the spinal cord leading to complete paralysis of the muscles that are supplied by that area. In abortive and nonparalyptic forms, recovery is normally complete. In Paralytic poliomyelitis 25% suffer permanent disability.
Possible Allopathic Treatment	Active immunization to prevent illness should be considered. Therapy will be symptomatic, ranging from bed rest to drug therapy.

Disease or Disorder	**RADICULOPATHY**
Description	Damage to or degeneration of a nerve root. Can be caused by a prolapsed invertebral disc causing sciatica or cervical osteoarthrosis. May cause brachial neuralgia.
Possible Allopathic Treatment	Rehabilitation, disorder management, patient education, exercise. Drugs include, aspirin, NSAIDs, Muscle relaxants, intra-articular corticosteroids, analgesics and tricyclic antidepressants. Surgery ie. Laminectomy, osteotomy and total joint replacement are considered when all other types of therapy have failed. Tens (transcutaneous electrical nerve stimulation) occasionally proves useful.

Terminology

Ganglia: a group of nerve cell bodies usually located outside the brain and the spinal cord.

Plexus: a network of interlacing nerves.

Spasticity: a state of sustained contraction of a muscle associated with an exaggeration of deep reflexes.

Synapse: the region of communication between neurons. A point at which an impulse passes from an axon of one neuron to a dendrite of the cell body of another.

The Endocrine System

Introduction

The endocrine or ductless glands pass their secretions or hormones directly into the blood stream. These hormone secreting glands are distributed throughout the body and include:

1. The pituitary,
2. The thyroid,
3. The parathyroids,
4. The thymus,
5. The supra-renal glands or adrenals,
6. Part of the pancreas and
7. Parts of the ovaries and testes.

Although these glands are separate, it is certain that there function is closely related to each other, because the health of the body is dependant upon, the correctly balanced output from the various glands that form this system.

The Pituitary Gland (Hypophysis)

This gland has been described as the leader of the endocrine orchestra. It consists of two lobes, anterior and posterior.

The anterior lobe secretes many hormones including the growth promoting **somato-tropic** hormone which controls the bones and muscles and in this way determines the overall size of the individual.

Over secretion of the hormone in children produces gigantism and under secretion creates "dwarfism". The anterior lobe also produces **gonadotropic** hormones for both male and female gonad activity. **Thyrotropic** hormones regulate the thyroid and **adrenocorticotropic** hormones regulate the adrenal cortex. It also produces **metabolic** hormones.

The posterior lobe produces two hormones:

1. **Oxytocin** which causes the uterine muscles to contract. It also causes the ducts of the mammary glands to contract and, in this way, helps to express the milk which the gland has secreted into the ducts.
2. The second hormone it produces is **vasopressin** which is an antidiuretic hormone. This has a direct effect on the tubules of the kidneys and increases the amount of fluid they absorb so that less urine is excreted.

It also contracts blood vessels in the heart and lungs and so raises the blood pressure. It is not certain whether these two hormones are actually manufactured in the posterior lobe or whether they are produced in the hypothalamus itself and passed down the stalk of the pituitary gland to be stored in the posterior lobe and liberated from there into the circulation.

The Thyroid

The right and left lobes of this gland lies on either side of the trachea united by the isthmus. Averages size of each lobe is one and a half inches long and three-quarters of an inch Across but these sizes may vary considerably. The secretion of this gland is thyroxine and tri-iodothyronine. Thyroxine controls the general metabolism. Both hormones contain iodine but thyronine is more active than thyroxin.

Under secretion of this hormone in children produces the cretin, that is a child who grown up a dwarf and an idiot whilst under secretion in adults results in a low metabolic rate.

Over secretion in adults gives rise to exophthalmic goitre and the metabolic rate is higher than usual. Such persons may eat well but burn up so much fuel that they remain thin, usually accompanied by a rapid pulse rate. This gland, therefore, has a profound influence on both mental and physical activity.

DIAGRAM OF GLANDS OF THE BODY

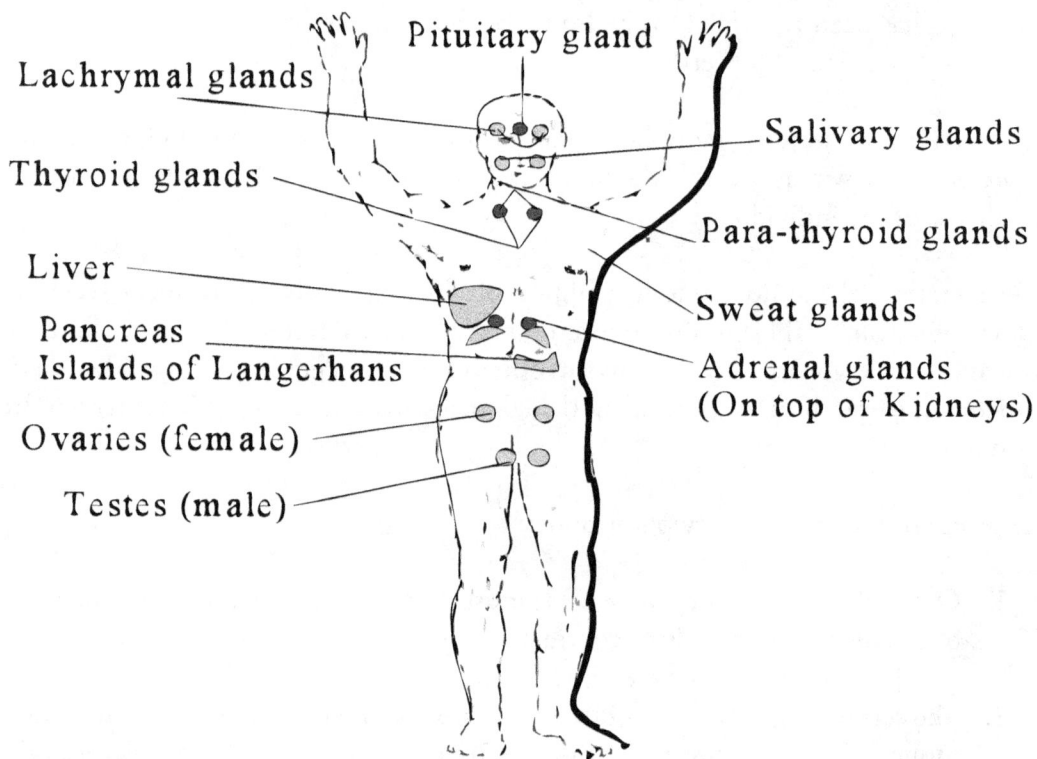

Pituitary gland

Lachrymal glands

Salivary glands

Thyroid glands

Para-thyroid glands

Liver

Sweat glands

Pancreas
Islands of Langerhans

Adrenal glands
(On top of Kidneys)

Ovaries (female)

Testes (male)

Endocrine System

Hypothalamus ← —— Input from eyes, thalmus, Limbic System. Adrenas

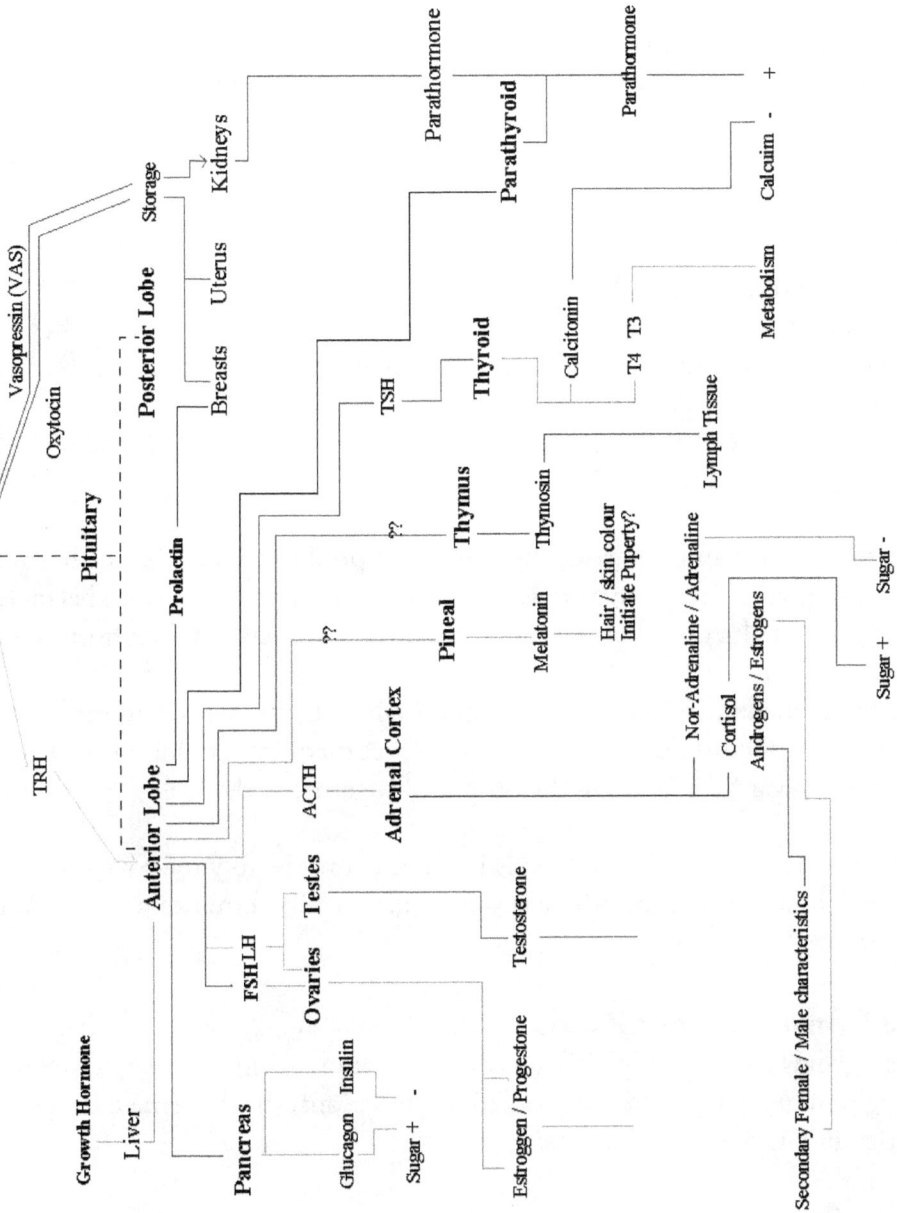

Vasopressin (VAS)

Oxytocin

TRH

Pituitary

Posterior Lobe

Storage → Kidneys

Uterus

Breasts

Parathormone

Parathyroid

Parathormone

Parathormone

Anterior Lobe

Prolactin

TSH

Thyroid

Calcitonin

T4 T3

Metabolism   Calcuim -   +

Thymus

Thymosin

Lymph Tissue

??

Pineal

Melatonin

Hair / skin colour
Initiate Puperty?

Nor-Adrenaline / Adrenaline

Cortisol

Androgens / Estrogens

Sugar +   Sugar -

Adrenal Cortex

ACTH

??

Growth Hormone

Liver

Pancreas

Glucagon   Insulin

Sugar +   -

FSH LH

Ovaries   **Testes**

Testosterone

Estrogen / Progestone

Secondary Female / Male charicteristics

107

The Parathyroid Gland

There are four of these glands in total, two on either side and lying behind the thyroid. Their secretion is parathormone, the function of which is to raise the blood calcium as well as maintain the balance of calcium and phosphorous in both the blood and bone structures. Under secretion gives rise to a condition known as tetany in which the muscles go into spasm, and over secretion causes calcium to be lost to the blood from the bones giving rise to softened bones, raised blood calcium and a marked depression of the nervous system.

The Thymus Gland

This gland lies in the lower part of the neck and attains a maximum size of about two and a half inches long. After puberty the thymus begins to atrophy so that in the adult only fibrous remnants are found. Its secretion is thought to act as a brake on the development of sex organs so that as the thymus atrophies, the sex organs develop.

The Suprarenal or Adrenal Glands

These are two in number, triangular in shape and yellow in colour. They lie one over each kidney. They are divided like the kidney into two parts:

1.　The cortex and
2.　The medulla.

The cortex is the outer part of the gland and produces a number of hormones called cortico-steroids. Their function is to control sodium and potassium balance, stimulate the storage of glucose, affect or supplement the production of sex hormones.

Whilst the medulla or inner layer produces adrenalin which is a powerful valvular constrictor, it raises the blood pressure by constriction of smaller blood vessels and it raises the blood sugar by increasing the output of sugar from the liver.

The amount of adrenalin secreted is increased considerably by excitement, fear, or anger, which has caused the adrenals sometimes to be referred to as the glands of fright and fight.

The Gonads or Sex Glands

These glands are naturally different in men from women because they serve different, though, in many respects, complimentary functions. In the female the gonads are the ovaries and in the male the testes.

- Female sex hormones are oestrogen and progesterone.
- The male sex hormone is testosterone.

Though each sex produces a small quantity of the opposite hormone. The female hormones are responsible for developing the rounded, feminine figure, breast growth, pubic and axillary hair and all the normal manifestations of femininity and reproduction. Male hormone is responsible for voice changes, increased muscle mass, development of hair on the body and face and the usual development of manliness.

Pancreas

The part of the pancreas which is the endocrine gland is called islets of langerhans and the hormone produced by this part of the pancreas in insulin which regulates the sugar level in the blood and the conversion of sugar into heat and energy.

To little insulin results in a disease known as diabetes mellitus. This is a disease which is strictly divided into one of juvenile onset, and is before the age of 25, and that which begins in maturity. It is a very common disease.

It is known that some half million people in the united kingdom suffer from it sufficiently to need treatment but it has been estimated that there are many more people in whom the disease exists but at a sub-treatment level.

Drs. Rankin and Best, succeeded in 1922 in keeping a diabetic dog alive in a Canadian laboratory. Since then, by injection and, more recently, surgery, it has been possible to contain this disease although the supplement of insulin is really a support treatment rather than a cure.

Additional Notes on the Hormone System

The quantities involved in the secretion of the various glands is quite minute, for example, the adrenal glands, which affect all the organs of the body, produce in a complete year not more that 1 gramme of hormones or three spoonfuls in an entire lifetime.

Some hormone deficiencies appear to be endemic, that is they are particularly prevalent in certain parts of the world and the best example probably is to be found in the diseases which affect the thyroid through lack of iodine. For example, endemic cretinism is common to the upper valleys of the Alps and the Himalayas where endemic goitre is also present, whilst the latter condition is to be found also in the region of the great lakes and the valley of saint Lawrence.

In all these areas, which are deficient in iodine or iodine containing foods, the authorities now usually take precautionary measures such as the provision of iodised salts in order to stop the development of the disease.

Terminology

Addison's Syndrome

This is due to adrenal, cortical insufficiency, and is characterised by hypotension, wasting, vomiting and muscular weakness.

Amenorrhoea

Absence of menstruation

Cushing's Syndrome

Caused by over secretion of the adreno-cortical hormones characterised by moon face, re-distribution of body fat, hypertension, muscular weakness and occasionally mental derangement.

Hyperthyroidism

This is known by a variety of names, thyrotoxicosis, toxic goitre and graves' disease or exophthalmic goitre. In hyperthyroidism many of the body's physical activities are subject to a speeding up whilst the opposite condition, Hypothyroidism is evidenced by a slowing down of the body's activities.

Lesion

An alteration of structure of functional capacity due to injury or disease.

Menopause (Climacteric)

Menopause literally means cessation of menses and refers to the period in the female development when the reproductive function comes to an end. Linked to a decline in the supply of hormone secretions by the ovaries. This is not a disease, it purely marks another stage in the female progression through life.

Premenstrual Tension

This is the syndrome of depression, irritability, bloating, swelling and restlessness that occurs for about one week before the onset of menstruation.

Steroid

This is the generic name given to various compounds of internal secretions including the sex hormones.

Syndrome

A group of symptoms and signs which when considered together characterises a disease.

The Respiratory System

Introduction

The respiratory system is responsible for taking in oxygen and giving off carbon dioxide and some water. It normally divides into the upper respiratory tract and the lower respiratory tract. The process of taking in air into the body is normally referred to as inspiration and getting rid of air from the body expiration.

DIAGRAM OF THE RESPIRATORY SYSTEM

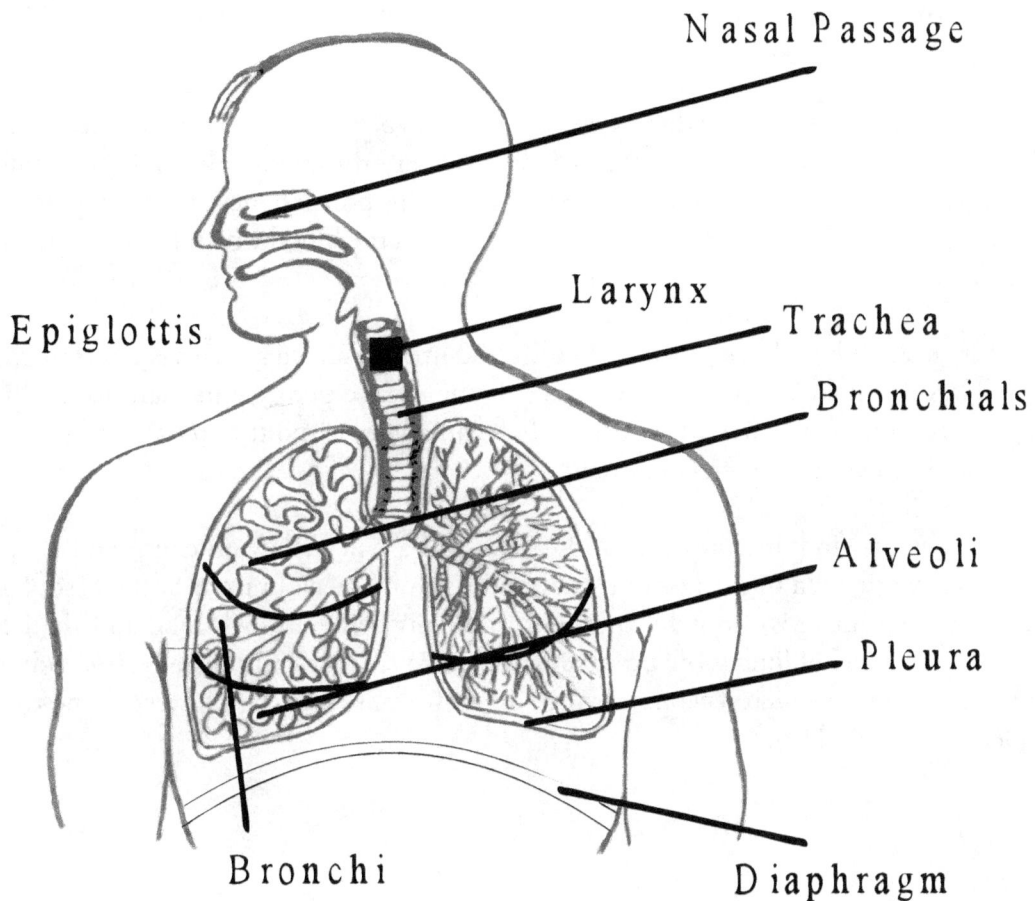

Nasal Passage

Larynx

Trachea

Bronchials

Epiglottis

Alveoli

Pleura

Bronchi

Diaphragm

The respiratory system consists of:

- The **upper respiratory tract**, which includes the nose, the mouth, the throat, the larynx and numerous sinus cavities in the head. Air brought in through the nose is filtered and warmed before passing down a tract into the lungs.
- The **lower respiratory tract.** Includes the trachea (or windpipe), the bronchi and the lungs themselves which contain bronchial tubes bronchioles and alveoli or air sacs.
- The **two lungs which are the principal organs of the respiratory system**, and are situated in the upper part of the thoracic cage. They are **inert organs**, which means, they do not work by themselves but they work by a **variation of atmospheric pressure** which is achieved by a muscular wall known as the **diaphragm**.

The contraction and relaxation of the diaphragm results in an alteration of atmospheric pressure within the lungs themselves. When the pressure is increased the air in the lungs rushes out this is called expiration. When the pressure is decreased the air rushes in which is called inspiration.

The inspired air, which contains oxygen, passes down into the billions of minute air chambers or air cells known as alveoli which have very thin walls. Around these walls are the capillaries of the pulmonary system. It is at this point that the fresh air gives off its oxygen to the blood and takes carbon dioxide from the blood which is then expelled with the expired air.

An average adult breathes something like 13,500 litres of air a day. This is not only the body's largest intake of any substance but it is the most urgently important to the life process. It is possible to live without food for many days, without water a few days but without air only for a very few minutes.

The trachea or windpipe measures about four and a half inches in length and is approximately one inch in diameter. It passes through the neck in front of the oesophagus branching into two bronchi, the right bronchus being an inch long and the left bronchus two inches long. The bronchi branch into smaller and smaller tubes ending in the bronchioles which have no cartilage in their walls and have clusters of the thin walled air sacs or alveoli.

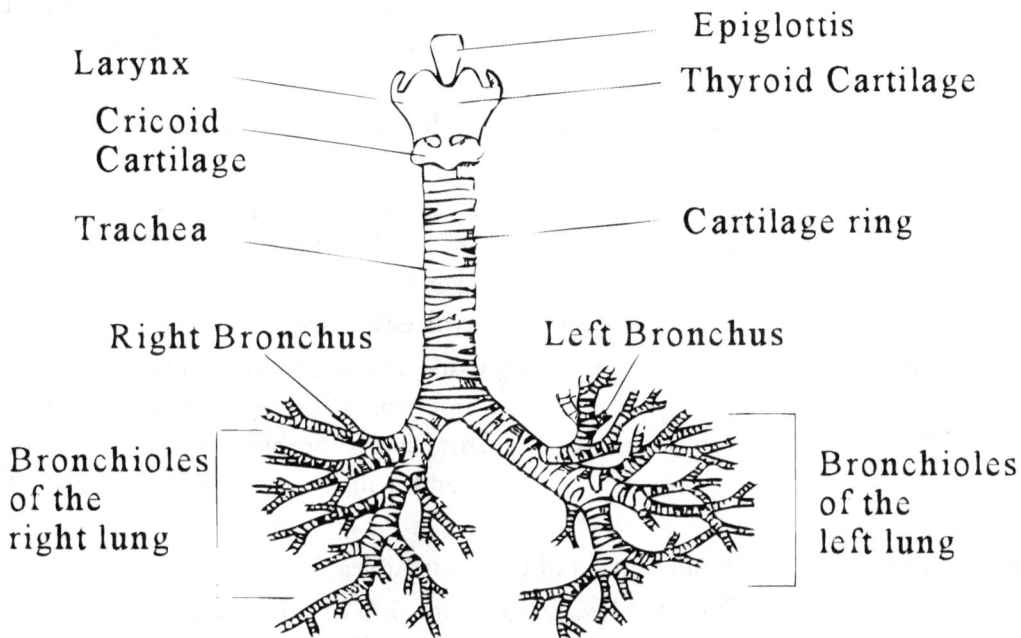

Epiglottis
Thyroid Cartilage
Larynx
Cricoid Cartilage
Trachea
Cartilage ring
Right Bronchus
Left Bronchus
Bronchioles of the right lung
Bronchioles of the left lung

The lungs are greyish in colour and are spongy in appearance. The right lung has three lobes divided into the, upper, middle and lower, and the left lung has two lobes divided into the upper and lower portions, because it must leave room for the heart.

The pleura is the serous membrane which covers the lungs.

The visceral layer is in close contact with the lung tissue and the parietal layer lines the chest wall, between these layers being the pleural cavity. In health it is a natural cavity because the two membranes are fluid lubricated on their opposing surfaces and slide easily over each other as the lungs expand and contract. Air going into the lungs follows the same throat passageway as food for a short distance but there is a ingenious trapdoor called the epiglottis which permits the passage of air to the lungs, but closes it when liquids are swallowed.

With each quiet breath cycle of an adult about on pint (.568 litres) of air flows into and out of the lungs. This is known as tidal air. If one continues to inhale at the end of a quiet inspiration almost an additional half pint (.284 litres) of complemental air can be forced into the lungs. Equally, if one continues to exhale at the end of exhalation, Almost a pint and a half (.852 litres) of supplemental air can be forced out of the lungs. However, there always remains in the lungs about one and a half pints (.852 litres) of air which is known as residual air and this cannot be forced out. The normal rate of inspiration and exhalation is about sixteen times a minute.

Some Conditions of the Respiratory System

Asthma	Difficulty in breathing due to contraction of the smooth muscle in the bronchioles. This gives a wheezing sound on respiration. There is a variety of causes, including allergy and infection
Atelectasis	Collapse of part of the lung, usually due to blockage of a small bronchus.
Bronchial Carcinoma	This is usually due to a cancer of the mucous membrane of a bronchus. It spreads to involve The local area of the lungs and the local lymph glands. Cancer can spread from other parts of the body to involve the lung.
Bronchiectasis	An area of the lung tissue and small bronchi that has been damaged by repeated infection, causing scarring and distortion of the tissues. This leads to problems with drainage of the normal secretions and a much increased likelihood of further infection.
Bronchiolitis	Inflammation of the bronchioles.
Bronchitis	This may be an acute infection of the bronchi that resolves with out leaving any damage. Chronic infections may occur when the mucosal cells are damaged and the normal drainage cannot take place. This means that the infection is unable to clear up completely. This occurs in two forms: acute bronchitis and chronic bronchitis.
Acute bronchitis	May result from inhaled materials, or it may be connected with another disease condition like influenza, measles, whooping cough.
Chronic bronchitis	Normally occurs at middle age or later and is four times more prevalent in men than in women. The disease can prove fatal and about 30000 deaths are recorded annually in Britain from this cause.
Emphysema	A condition of the alveoli in which their walls are damaged and broken down. This reduces the Elasticity of the lungs and also the surface area over which the exchange of gasses may take place.
Laryngitis	Hoarseness of the voice due to inflammation of the vocal cords and the larynx.
Pharyngitis	Inflammation of the pharynx at the back of the throat.
Pleurisy	Inflammation of the pleura. Practically any disease that causes inflammation of the lungs may result in pleurisy. The pleura becomes inflamed and fluids accumulate in the interspace.

Pneumoconiosis	Damage to the lung tissue from inhalation of dust. The term pneumoconiosis indicates a lung condition due to inflammation by minute particles of mineral dusts. It is often called miner's disease or miner's lung due to it prevalence amongst this body of workers. There are, however, a number of other occupations where find dust is a hazard and workers who are exposed to a high concentration of silica dust may develop a variation of the disease known as pneumosilicosis. Precautionary measures like the wearing of masks is helping to reduce the incidence of this disease.
Pneumonia	Infection of the alveoli causing impairment of breathing. This may either have spread from the bronchi, bronchopneumonia, or occur in the whole of the lobe, lobar pneumonia.
Pneumothorax	Air in the pleural cavity that may occur after the rupture of a lung.
Rhinitis	Inflammation of the nasal passages. This may be due to allergy, hay fever.
Sinusitis	Inflammation of the mucous lining of the sinuses.
Tonsillitis	Infection of the tonsils.
Tracheitis	Infection of the trachea.
Tuberculosis	An infection of the lung caused by an organism of tuberculosis. This is a chronic illness that used to be a common occurrence. This disease known as tuberculosis has been with us for thousands of years. Centuries before Christ, it was called phthisis, this is a Greek work meaning to consume slowly or waste away. This explains the familiar word for this disease, Consumption. It was in 1882 that a German Bacteriologist, Robert Koch, discovered that tuberculosis was caused by a long shaped bacterium called tubercle bacillus.

The Genito-Urinary and Kidney System and the Reproductive System

The Genito-Urinary System

Many anatomical textbooks divide the genito-urinary systems into the reproductive system and the excretive system. But because a number of the organs involved are common to both systems, for the purpose of this course we will treat them under this one heading.

The principal organs involved in this dual system are:
- the ovaries
- fallopian tubes
- uterus
- testes
- urethra
- ureter
- the urinary bladder

There are a number of smaller accessory organs involved and these will be dealt with appropriately in due course.

Diagram of the section of the female pelvic cavity

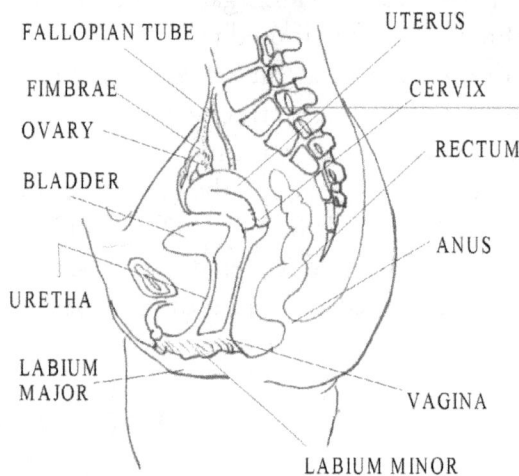

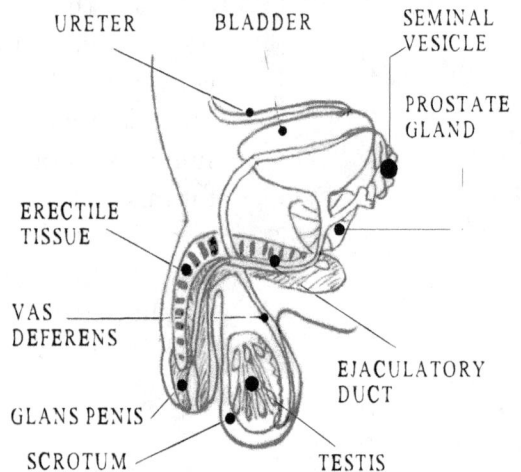

Diagram of the section of the male pelvic cavity

FALLOPIAN TUBE
FIMBRAE
OVARY
BLADDER
URETHA
LABIUM MAJOR
UTERUS
CERVIX
RECTUM
ANUS
VAGINA
LABIUM MINOR

URETER
BLADDER
SEMINAL VESICLE
PROSTATE GLAND
ERECTILE TISSUE
VAS DEFERENS
GLANS PENIS
SCROTUM
EJACULATORY DUCT
TESTIS

First we have the right and left ovaries in the female anatomy. These are quite small and are about the size of an almond.

They consist of masses of very small sacs known as the "ovarian follicles" and each follicle contains an "ovum".

The ovaries would appear to have two principal functions.
1. To develop the ova and expel one at approximately twenty-eight day intervals during the reproductive life.
2. To produce hormones (oestrogen and progesterone) which influence secondary sex characteristics and control changes in the uterus during the menstrual cycle.

The fallopian tubes - Sometimes referred to as the uterine tubes, are about four inches long and their function is to transport the ova from the ovaries to the uterus.

The uterus is a muscular organ approximately pear-shaped, about three inches long by two inches wide and inch thick. It is positioned in the centre of the pelvis with the bladder in front and the rectum behind.

For the purpose of description, it is normally divided into three parts:
1. The fundus being the broad, upper end.
2. The body is the central part.
3. The cervix is about an inch long and is the neck which projects into the vagina.

The vagina is the muscular canal which connects the above organs to the external body at the point collectively known as the vulva which includes the clitoris which is a small, sensitive organ containing erectile tissue corresponding to the male penis.

The male genital organs are comparatively simple in comparison to the female genital organs. The principal organs are the testes or testicles which are the essential male reproduction glands.

The scrotum which is a pouch-like organ containing the testes and the penis which is suspended in front of the scrotum.

The kidneys are two bean shaped organs, approximately four inches long, two inches wide and one inch thick. They are positioned against the posterior wall at or above the normal waistline. quite often the right kidney is slightly lower than the left kidney.

The kidneys consist of three principal parts:
1. The cortex or outer layer which is light brown in colour.
2. The middle portion which is inside and is dark brown in colour.
3. The pelvis which is the hollow, inner portion from which the Ureters open.

The function of the kidneys is to separate certain waste products from the blood and this renal function helps maintain the blood at a constant level of composition despite the great variation in diet and fluid intake.

DIAGRAM OF SECTION OF KIDNEY

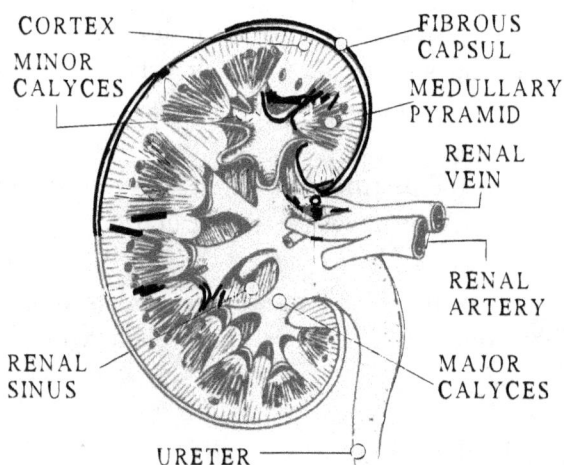

CORTEX
MINOR CALYCES
FIBROUS CAPSUL
MEDULLARY PYRAMID
RENAL VEIN
RENAL ARTERY
RENAL SINUS
MAJOR CALYCES
URETER

As blood circulates in the kidneys a large quantity of water, salts, urea and glucose are filtered into the capsules of bowman and from there into the convoluted tubules. From here all the glucose, most of the water and salts and some of the urea are returned to the blood vessels.

The remainder passes via the calyces into the kidney pelvis as urine. It is estimated that 33-40 gallons (150-181 litres) of fluid are processed by the kidneys each day but only about 2 and half pints (1.42 litres) of this leave the body as urine.

DIAGRAM OF INTERNAL KIDNEY STRUCTURE

The ureters are two fine muscular tubes, ten to twelve inches long, which carry the urine from the kidney pelvis to the bladder. This is a very elastic muscular sac lying immediately behind the symphysis pubis.

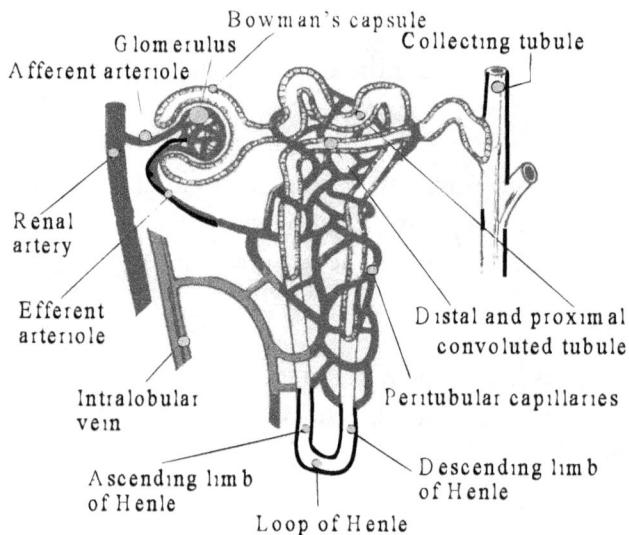

Bowman's capsule
Glomerulus
Afferent arteriole
Collecting tubule
Renal artery
Efferent arteriole
Distal and proximal convoluted tubule
Intralobular vein
Peritubular capillaries
Ascending limb of Henle
Descending limb of Henle
Loop of Henle

The urethra is a narrow muscular tube passing from the bladder to the exterior of the body. The female urethra is 1 to 1 ½ inches long and the male urethra 6 -8 inches long.

In the male, the urethra is the common passage for both urine and the semen or reproductive fluid. Also, in the male, it passes through a gland known as the prostate gland which is about the size and shape of a chestnut. It surrounds the neck of the bladder and tends to enlarge after middle life when it may, by projecting into the bladder produce urine retention.

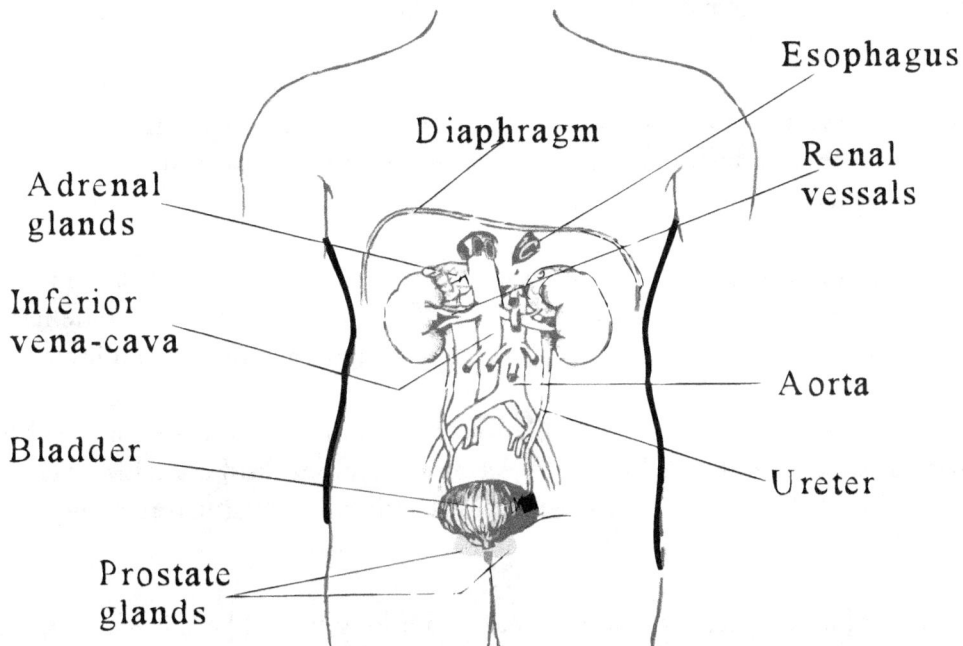

Additional notes on the functions of the genito-urinary system

The average composition of urine is 96% water and 4% solid. Of the 4% , 2% represents urea and 2% salts.

The 2% urea compares with 0.04% urea in blood plasma so it will be seen that concentration has been increased some fifty times by the work of the kidneys.

The salts consist mostly of sodium chloride, phosphates and sulphates produced partly from the presence of these salts in protein foods.

These salts have either to be reabsorbed or got rid of by the kidneys in sufficient quantities to keep the blood at its normal reaction and balance.

The urine also contains traces of a number of other substances, all of which combine to produce, in the urine, a reasonable pattern of the state of the body itself.

Its analysis indicates a number of physiological states including the amount of alcohol in the body, whether a female is pregnant or not and whether a person has diabetes.

It is estimated that, at birth, there are some thirty thousand ova or eggs in a female child. No fresh ova are formed after birth but, during the reproductive female life,

That is commencing between ten and sixteen years of age and concluding between 45 and 55 years of age, these ova develop within the follicles or sacs in which they are embedded and progressively come nearer to the surface of the ovary where they mature and increase in size.

Then, about every 28 days, one of these follicles bursts and the ovum it contains, together with the fluid surrounding it, is expelled into the fallopian tubes and then into the uterus where it may or may not be fertilised.

If the ovum is fertilised by a male Reproductive cell or **spermatozoon** it then attaches itself to the uterine wall and there develops. If the ovum does not become fertilised within a few days it is cast off and the process termed **menstruation** is initiated.

The spermatozoa which are responsible for fertilisation are contained in a substance known as seminal fluid. An average ejection of **seminal fluid** contains several hundred million of these mobile units which look rather like a miniature elongated tadpole, and 1/500th of an inch in light.

It consists of a headpiece, a middle piece and a long whip-like tail which gives the spermatozoon its mobility. The single fertilized ovum soon becomes many cells which develop in a bag of membranes which soon fill the uterine cavity. At one part of this sac, at the point where the ovum first embedded itself in the uterine wall, the **placenta** or afterbirth develops.

The umbilical cord contains blood vessels and runs from the navel of the foetus to the placenta. The placenta receives the mother's blood from the wall of the uterus and the infant's blood via the umbilical cord so that, at no stage, does the mother's blood pass directly into the child. It is through the placenta that the child's blood is able to absorb food, oxygen and water from the mother and, in turn, give off its waste products.

The skin is, of course, an organ very closely connected with the excretal system but, as it is a multipurpose organ, it is dealt with later on in the course.

The mammary glands or breast

These are accessories to the female reproductive organs and secrete milk during the period of lactation. They enlarge at puberty, increase in size during pregnancy and atrophy in old age.

The breast consists of mammary gland substance or **alveolar** tissue arranged in lobes and separated by connective and fatty tissues. Each lobule consists of a cluster of alveoli opening into lactiferous ducts which unite with other ducts to form large ducts which terminate in the excretory ducts. As these ducts near the nipple expand they create a reservoir for the milk, thus being called the **lactiferous sinuses**.

The breast also contains a considerable quantity of fat which lies in the tissues of the breast and also in between the lobes. It also contains numerous lymphatic vessels which commence as tiny plexuses which unite to form larger vessels which eventually pass mainly to the lymph node in the axilla. The nipple is surrounded by a darker coloured area known as the mammary areola.

The breasts are very greatly influenced by hormone activity. Hyper-secretion of the thyroid can lead to atrophy of the breasts whilst hypo-secretion can be the cause of too greatly developed breasts. Both the ovarian hormones influence the condition and appearance of the breast whilst the pituitary hormone **prolactin** starts lactation at the end of pregnancy.

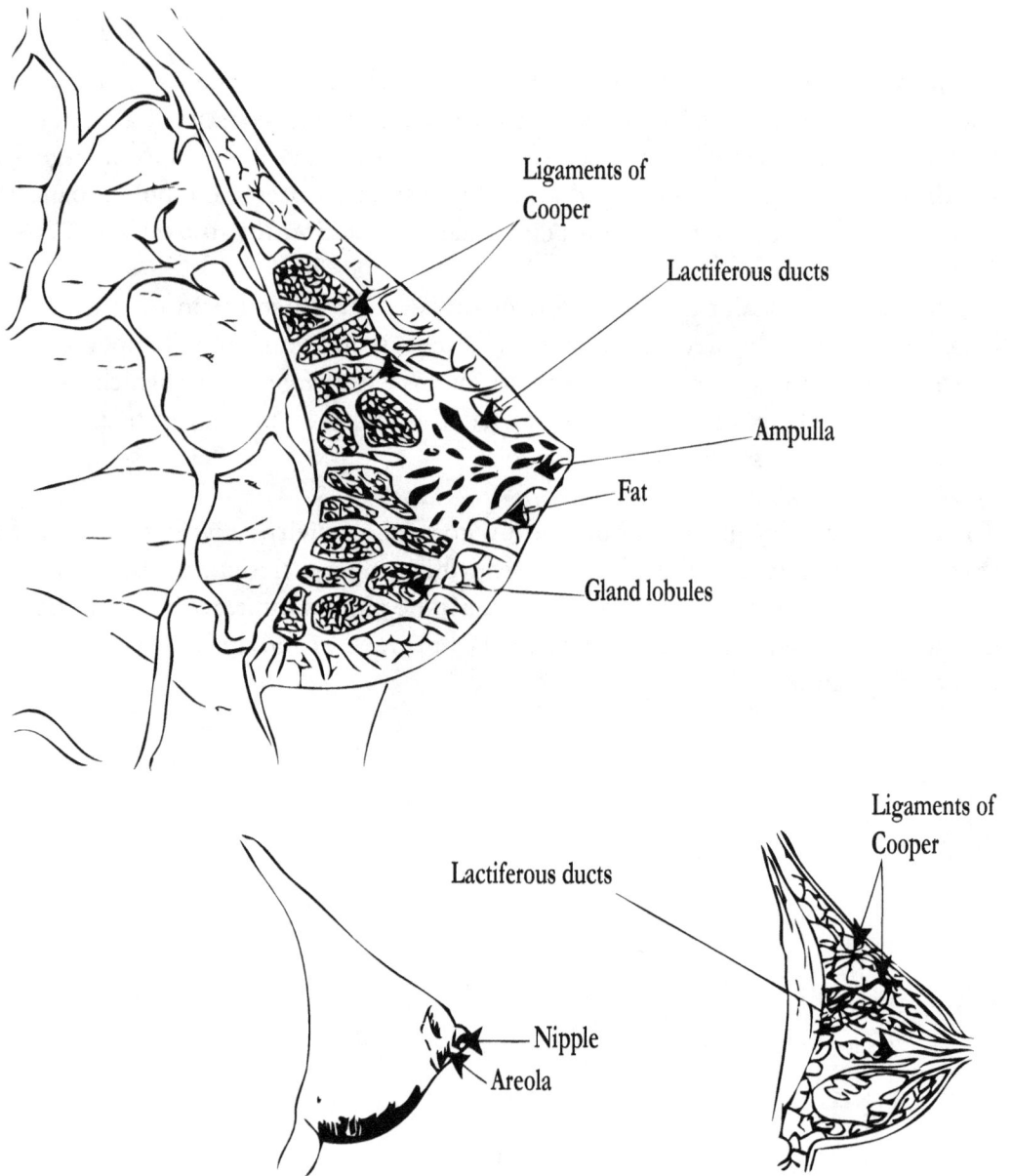

Ligaments of Cooper

Lactiferous ducts

Ampulla

Fat

Gland lobules

Lactiferous ducts

Nipple

Areola

Ligaments of Cooper

Some Condition of the Genito-Urinary System

Cervicitis — This is an infection of the cervix that is, the neck of the uterus, and is reasonable common. It may be due to gonorrhoea, syphilis or a specific infection.

Cystitis — Cystitis is inflammation of the bladder. A condition especially common in women. This is due to the fact that the urethra in women is a very short one and is a pathway to invasion by infecting organisms.

Kidney stones — Stones in the kidney are quite common and precipitate out of the urine, which has already been explained as a complex solution of many substances. Surgical operations for the removal of stones have a very long history. The Greek Doctor Hippocrates, admonished fellow physicians not to cut out stones but to leave it to the specialist. Whilst, nearer to our time, Samuel Pepys describes his own operation for the cutting of stone on march 26th 1658. He notes in his diary that he spent twenty four shillings "for a case to keep my stone that I was cut of".

Nephritis or Bright's disease — This was first described by Dr. Richard Bright of London, England, in 1827. The single disease which he diagnosed has now been subdivided into a number of conditions which may, in a broader way, be called nephritis; An inflammation of the kidney not resulting from infection in the kidney.

Calculus — Stone e.g. Renal calculus; Stone in the kidney

Catheter — A hollow tube for introduction into a cavity through a narrow canal for the purpose of discharging fluid from a cavity. For instance, passing water from the bladder for the relief of urinary retention.

Dysmenorrhoea — Painful menstruation

Ectopic — Gestation development of the ovum in a fallopian tube instead of the uterus

Enuresis — The involuntary discharge of urine.

Foetus — The unborn child dating from the end of the third month until birth

Intra-uterine — Within the uterus. Relating to conditions which occurred before birth

Menopause — Also called climacteric. The physiological cessation of menstruation.

Micturition — The act of passing urine.

Mechanical methods of contraception

1. The natural or rhythm method which requires restricting intercourse to the non-fertile period of the woman which only lasts a few days after ovulation. Taking the body temperature assists in determining the exact date of ovulation.

2. Birth control pill taken orally contains a mixture of oestrogen and progesterone-like hormones which fool the body into thinking it is already pregnant via the endocrine system.

3. Surgical intervention involves a vasectomy for men and tubal ligation for women. This is where the reproductive tubes are simply cut and tied.

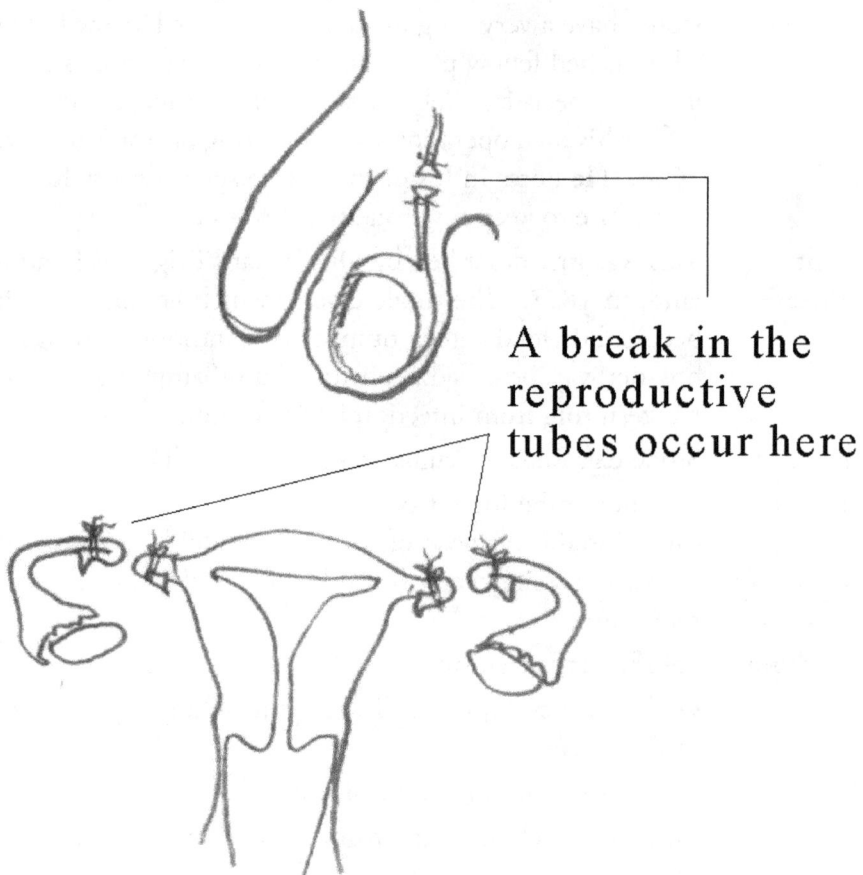

A break in the reproductive tubes occur here

The Skin

The Cell

Our bodies are composed of cells, tissues organs and systems. Each of these could not exist without the other, as they depend upon one another on an interactive level for health and communication. The levels of organization are as follows:

Cells build **tissues**
Tissues form **organs**
Organs create a **body system**
Body systems form an **organism** ie. humanity.

The cell is the smallest, basic structural unit of the skin, yet it is responsible for the body's metabolism and reproduction. It is the fundamental basic unit of all living matter. Although every part of the body is made up of cells, each cell is different in size, shape and structure, according to its function. Therefore, each cell is specialized to allow the body to carry out its vital functions. For example, blood cells transport oxygen through the blood vessels and remove carbon dioxide. Nerve cells communicate and co-ordinate the body systems, muscle cells provide movement of the limbs, and bone cells support the body.

Function of the Cell
1. To create and repair all parts of the body.
2. To help blood circulation by carrying food to the blood and waste matter away from the blood.
3. To control all body functions.
4. Certain cells are capable of reproduction.
5. The cell is capable of metabolism (absorbing and consuming nutrients to form energy)

Structure of the Cell
It is estimated that the human body consists of over 75 trillion cells. The individual size of a person is not determined by the cell size, but rather the number of cells an individual possesses. The cell has three main components:

1. Cell (plasma) Membrane - this forms the outer boundary
2. Cytoplasm -this is found inside the cell membrane and surrounds the nucleus and other organelles.
3. Organelles - these are specialized structures within the cell that carry out specific functions

1. **Cell Membrane**

This is a very thin porous skin that is composed of several layers of molecules. The inner and outer layers are composed of protein molecules. Between this is a double layer of fat molecules (lipids). These lipids provide a barrier between water soluble materials both inside and outside the cell and protects the cell from the external environment.

2. **Cytoplasm**

Is the living liquid matter that is found between the cell membrane and the nucleus. It contains 75 - 90% water plus organic and inorganic compounds. Its consistency is slightly thicker than water which holds the organelles in suspension.

3. **Organelles**

These are specialized structures the perform specific cellular functions. Some of these include:

Centrosomes	Which are important in cell reproduction.
Mitochondria	Which contain enzymes that provide the cell with energy and also is responsible for cell respiration.
Vacuoles	Are fluid filled bubbles found in the cytoplasm that contain food material or waste supplies
Ribosomes	Make proteins for the cells use.
Lysosomes	Manufacture digestive enzymes that break down large molecules into smaller ones, so these products can be converted into other necessary chemicals and substances.

4. **Nucleus**

Is found in the centre of the cell.
- It is enclosed by a nuclear membrane, which is a thin porous layer of tissue.
- It is responsible for cell reproduction and cell metabolism.
- It contains one or more small bodies called **nucleoli** which float freely inside the nucleus, which are responsible for the synthesis of RNA.

The nucleus is the largest organelle of the cell. It is the control centre of the cell because it is responsible for all metabolic functions and plays an important part in cell reproduction.

Both **RNA** and **DNA** (*ribonucleic acid and deoxyribonucleic acid*) are found within the nucleus.

DNA is a long spiral-shaped molecule (*similar in appearance to a twisted ladder*) that has the ability to duplicate itself, therefore passing on all necessary information to any new cells, DNA is also the building blocks of chromosomes.

The DNA contains genes, these genes contain an individuals' hereditary information or genetic code.

Each human chromosome (*there is 46 in total*) contains 20,000 genes. It is very easy to understand why no two people in the world are identical, taking into consideration the vast number of possible genetic combinations.

Cell Metabolism

Metabolism: Is the chemical change that absorbed foods undergo in the body cells.

There are two forms of metabolism:

1. **Anabolism**   The building up of cellular tissues. During anabolism, the cells absorb water, food, and oxygen for growth, reproduction and repair. This is considered the constructive part of metabolism.

2. **Catabolism**   Breaks down cellular tissue. During catabolism, the cells consume what they have absorbed to perform specialized functions, such as muscular effort, secretion, or digestion. This is considered the destructive part of metabolism.

Metabolism goes on continually inside the cell. A simpler example of metabolism is a people require nutrients and food for growth, repair and maintenance of a healthy body - this is anabolism because we are feeding the body to build up cellular tissue.

Then, during physical activity catabolism takes place because the cells are consuming or using up the nutrients they have absorbed by breaking down cellular tissue.

Cell Reproduction

Cells reproduce by division called **mitosis**. This enables growth and repair of tissue. It is the replication of a cell, into two daughter cells, resulting in both of these newly formed cells being identical to the original cell. Mitosis assures that any newly formed cells contain the same number and composition of chromosomes.

Mitosis is generally a process that is controlled. Therefore, the replication process will be dormant when there is no need for additional cells to be produced. Occasionally, the control is lost and the cells continuously undergo division, which lead to the formation of tumours. Tumours are either benign (*do not spread*) or malignant (*cancerous and do spread throughout the body*).

Cell Permeability

All materials that enter or exit the cell (ie: food, water, oxygen, etc.) must pass through this membrane. Only certain molecules are allowed to enter and exit. This is called selectively permeable.

The permeability of the cell will depend on four different things:

1. Size of molecules - water and amino acids are allowed easy access in and out of the cell, while larger protein molecules are restricted due to their size. (Essential oils are very minute in their molecular structure.)
2. Solubility in lipids - the cell membrane is composed of 50% of lipids. Therefore the natural path of penetration of the skin is through lipids. If the substance is soluble in lipids, entrance to the cell will be gained. (Essential oils are soluble in lipids.)
3. Presence of a carrier molecule - there are specialized molecules that are "carriers". Their sole purpose is to carry certain substances into the cell. Glucose requires a carrier in order to gain entry into a cell.
4. Charge of an ion - the skin is capable of ionization. In other words, if the charge of an ion (substance) is an opposite charge to that of the cell membrane, then the ion will be attracted into the cell. Similar charges will result in a "repelling" of each other.

The Skin

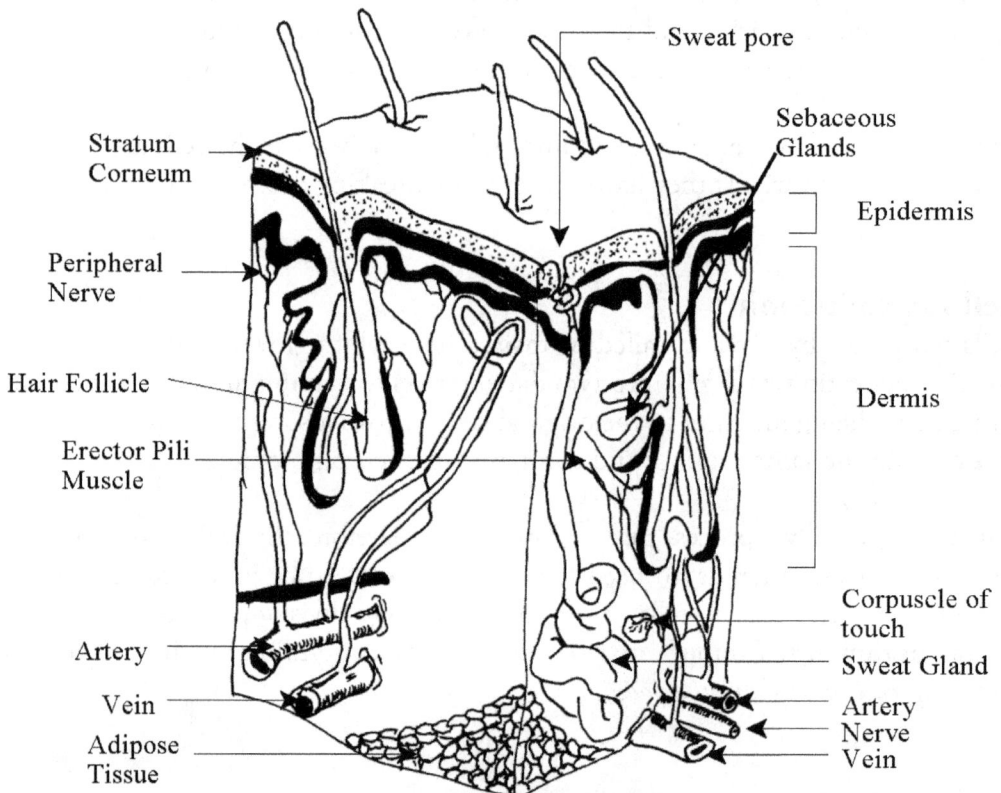

128

The skin and its derivatives (hair, glands and nails) constitute the integumentary system. It is the largest organ of the body. It makes up 15% of our total body weight. The skin is constantly shedding and replacing itself through new cells from below. It varies in thickness in different areas of the body. The thickest areas are the palms of the hands and the soles of the feet while the thinnest areas are the eyelids and the lips.

The skin provides a protective barrier that helps prevent damage occurring to internal organs from traumas as well as substances like chemicals, pollution, bacterial invasion, microorganisms and ultraviolet radiation.

Vitamin D is metabolized in the skin. This is an important ingredient in the formation of bone. The skin is able to absorb and excrete. It can move substances such as chemicals and drugs into the skin. Not only are they able to gain easy access into the skin (refer to the factors of permeability), but they are further absorbed into the bloodstream and carried throughout the body to assist in healing.

Temperature regulation of the skin occurs through specialized receptor cells called thermoreceptors. They will react to heat by inducing sweating, or in cold situations, they will create shivering to induce warmth. Touch receptors called Meissners corpuscles are located all over the skin.

The Epidermis

The epidermis is the outermost part of the skin that is composed of different layers. It is devoid of blood vessels, and nerves.

The top layer is known as the stratum corneum and is composed of tightly packed dead cells call keratin. This is the layer that is visible to us. These cells are rubbed off during the normal wear and tear of the skin.

Beneath the stratum corneum lies the stratum germinativum or granular layer. Then we have the squamous cell layer. This layer contains the most abundant cells of the skin. Then finally we have the stratum germinativum or the basal cell layer. This is the lowest layer of the epidermis and it sits upon the dermal layer. It contains specialized cells called melanocytes which produce a pigment called melanin, which is responsible for the colouring of the skin. The more active the cells, the greater the quantity of melanin produced, hence the varied colours in different races of people. Melanin protects the skin from sun damage and the darker your natural skin colour is, the greater protection you have from the sun. Sun tans do not protect your skin, remember to use a sun-screen for protection.

New cellular growth begins in the basal layer. As this growth takes place, the cells move upwards towards to stratum corneum and gradually keratinize (die) during this process. Once they have reached the upper layer, they have died. This cycle takes approximately 28 days for completion. The cells spend approximately 2 weeks moving towards the stratum corneum and then another 2 weeks before they are shed.

Dermis

The dermis, varies in thickness from 1-4mm, is composed of dense fibrous tissue that provides for strength, extensibility and elasticity. The dermis is the underlying, or inner layer of the skin. It is a highly vascular and sensitive layer of connective tissue, that is made up of three types of fibres. These are called collagen, elastin and reticulum.

The dermis contains blood vessels, lymph vessels, nerves, sweat and oil glands, hair follicles, erector pili muscles and papillae.

The nerve endings in the skin are responsible for the feelings of sensation. For the therapist it is the hands which are the most important communicating role. Touch is fundamental to the development of a healthy human being, and touch deprivation in the early stages of life is known to inhibit the emotional and physical growth of a child. It is important to remember that the need to be touched in a caring way does not stop in childhood. The power of a caring touch and its ability to heal, share empathy and to comfort is vital in the holistic field. Touch can be remarkably effective in the reduction of pain, lowering of blood pressure, and control of nervous irritability.

The main objective of touch is to soothe and to provide a comforting connection that is calming and allows the powerful healing mechanisms of the body to function.

Collagen and elastin fibres are abundant in the dermis. Collagen makes up about 70% of the dermis. It gives structure and support for cells and blood vessels, allows for stretching and contraction of the skin, provides strength and aids in the healing of wounds. Reticulum fibres help link bundles of collagen together while the protein called elastin, gives elasticity and pliability to the dermis..

Moisture is important in keeping this network supple. It is the skins collagen condition and not the facial muscles that causes the lines and the wrinkles. Cells within the dermis called **fibroblasts** are responsible for producing the collagen and elastin fibres. As we age, the fibroblasts decrease their production ability.

Subcutaneous Tissue

This is a fatty layer found below the dermis, also referred to as **adipose** tissue. It attaches the skin to the underlying tissue and organs. It is thicker than the dermis but is absent in areas such as the eyelids, nipples, and shins. It serves as a heat insulator and stores fats for use as energy, all while acting as a shock absorber to protect the internal organs from trauma. It varies in thickness according to age and health of the individual. It gives smoothness and contour to the body and is distributed differently in males and females.

Penetration Routes of the Skin

The skin, up until the 1960's, was believed to be a barrier and able to resist any form of substance penetrating into the body. Today however, it is an effective way to deliver essential oils and even drugs into the skin and the body. This route is known as transdermal penetration. The substances will be carried through the skin and into the body's tissue and into the blood stream. It must be realized that the size of the molecule is a deciding factor. Some molecules are too large to penetrate in.

Essential oils on the other hand are extremely minute and are able to penetrate the skin (*via the hair follicles and the sweat glands*). It is thought that they can also permeate between the skin cells (*intercellularly*) through the lipids (*saturated fats*) and, as essential oils are soluble in fat, they can enter the dermis this way. Once there, they can reach the blood capillaries and lymph, from where they are transported around the body by the circulatory system. Urine tested an hour after applying essential oil on the back of the hand has been found to contain the oil applied. This is just one of many tests which backs up this finding.

The four different ways the substances can penetrate the skin are as follows:
1. The hair follicle - Substances that are small enough are able to enter into the hair follicle and into the capillaries and lymph that are connected to them.
2. The sebaceous/sudoriferous glands - These provide a pathway for substances, similar to the hair follicle.
3. The transcellular route - This route is through the actual skin cells, eventually making their way into the dermal layer and into the blood and lymph systems.
4. The intercellular route - This route is the best vehicle for penetration of the epidermis. The lipids (*fatty fluids*) found in the intercellular cement (*in between the cells*) will accept substances in that are similar in composition. (*Remember in the study of cells that solubility in lipids allows for permeability.*)

Other Factors that Influence Penetration

An excessively thick stratum corneum will hinder product penetration. This is obvious due to the large build up of dead skin cells. People who suffer from an oily skin will not have as in-depth penetration as well. The excess sebum will slow down any absorbency. Temperature of the skin will also influence penetration. A warm skin is generally more accepting due to the fact that heat causes activity within the molecules thereby increasing cellular activity. Also remember that hot or warm temperatures (*steam, compresses, etc.*) will dilate the follicles while cool or cold temperatures will constrict the follicle.

Aging of the Skin

The natural process of wear and tear naturally affects the skin. At puberty, adjustment to increased hormonal activity usually manifests itself on the skin of the adolescent. By the age of twenty, the systems of the body should be fully grown and developed. If the body is healthy, the skin should be at its optimum, glowing, firm and without lines or wrinkles.

Aging of the skin begins at the age of 25, and becomes most noticeable in the mid-forties. It is from the mid-thirties that the facial skin starts to lose its firmness and fine lines and wrinkles appear and, if there is any loss of muscle tone, there is sagging of the skin.

In the forties and the fifties the lines of expression and furrows in the forehead deepen and loss of muscle tone will cause further sagging of the skin on the cheeks and neck. This is largely due to subcutaneous fat and connective tissue losing its elasticity and becoming less firm. The papillary (*upper*) layer of the dermis flattens out so that the skin is finer and thinner. During menopause the activity of the sebaceous glands is reduced and the skin becomes drier.

As we grow older the cells' capacity to reproduce, grow and renew themselves decreases; therefore, in the older person cell regeneration slows down due to lack of nourishment as the arterioles thicken and venous circulation is impaired. Over the years the sebaceous glands and hair follicles atrophy and there is general loss of colour and thinning of the hair. More keratin is produced and the skin becomes dry and looks dull. Skin tags may appear and age spots can occur, especially on the back of the hands.

This process is affected by environmental conditions (*sunlight, weather, chemical irritants*) as well as by intrinsic (*hereditary*) conditions.

To Summarize the Changes Within the Skin Due to Aging
1. breakdown of collagen and elastin fibres, resulting in wrinkles and sagging
2. loss of subcutaneous tissue, resulting in sensitivity to temperature changes
3. atrophy (wasting away) of the sebaceous glands, causing dry skin
4. decrease in melanin production (*this is a pigment that is found in the hair, skin and eyes that serves as a means of protection by screening out any harmful ultraviolet rays*) that results in grey hair, and pigmentation (age) spots.

The Effect of Water on the Skin
The skin is 50 to 75% water. This is maintained by the secretion of sebum which creates a protective coating on the surface of the skin. This layer of oil that is produced slows down the evaporation of water in the skin and prevents any excess moisture from penetrating into it. If the natural oils are removed, or lost, the protection is lost. This is why when you are dealing with a dry skin type, you are generally finding dehydration present as well.

Effects of Ultraviolet Radiation on the Skin
Ultraviolet radiation, whether its from the sun or a sun-bed, produces damaging changes within the skin. This damage is cumulative, over a period of many years, although one days overexposure can result in short term changes such as a severe sunburn. Long-term damage caused by excessive ultraviolet radiation results in increased wrinkling, loss of elasticity, liver spots, a "leather type" skin appearance, and skin cancer.

Unfortunately these effects do not appear for years, hence, uninformed teenagers and young adults accelerate the skins aging and open the door to skin cancers that may arise later in life. Tanning beds can increase these potential problems.

Skin cancers are the most common form of cancer. Most are curable through surgery. These are called Basil Cell Carcinoma and Squamous Cell Carcinoma. However, Melanoma Cancer of the Melanocytes spreads rapidly and if it is not detected in the early stages, it will result in death in 45% of the cases.

You can reduce the risks of overexposure to ultraviolet radiation by limiting the time in the sunlight between the hours of 11 am to 3 PM and by using a sun-screen with a minimum SPF 15. Remember that sun-screens are not cumulative. If you have applied an SPF 8 and then applied a SPF4 on top, you still only have a maximum SPF8 coverage. Also remember that sun-screens should be applied daily to any exposed area of the body irregardless of the season.

The Glands of the Skin

There are two types of duct glands found within the skin:
    1) the **sudoriferous** or sweat glands
    2) the **sebaceous** or oil glands

1)   There are approximately two million sudoriferous glands in the body. They consist of a coiled base and a tube-like duct which ends at the skins surface to form the sweat pore. Almost all parts of the body are supplied with these glands. These glands are under the control of the sympathetic nervous system regulate the body temperature and help to eliminate waste products from the body. They are richly supplied with blood vessels.

    There are two types of sweat glands:
    1. **apocrine glands** - these have a duct that empties the secretion into the hair follicle. They occur primarily in the axillary (underarm) and the pubic regions. They produce a more fatty type of secretion than the eccrine glands. Breakdown of the secretion by bacteria leads to body odour under the arms and in the groin. Their activity is increased by heat, exercise, emotions and drugs of a certain type.
    2. **eccrine glands** - most of the glands are of this type. These are found all over the body and they secrete their secretions onto the surface of the skin. Practically all parts of the body are supplied with sweat glands, but they are more numerous on the palms of the hands, soles of the feet, forehead and the underarms.

    The sudoriferous glands secrete perspiration or sweat. This is a mixture of water, salt, urea, uric acid, ammonia, amino acids, simple sugars and vitamins. It also has anti-bacterial substances which help the body's defence against infection.

2) The oil glands are small saccular, glandular organs. They secrete sebum out of their glands, through the hair follicle and then onto the surface of the skin. This sebum lubricates the skin and helps to prevent the evaporation of moisture. Sebum is a mixture of fats and lipids. Some of the substances found within this mixture are triglycerides, sphingolipids, glycolipids, phospholipids, cholesterol, ceramides, fatty acids and waxes. Sebaceous glands are found over the entire skin surface, excluding the palms of the hands and the soles of the feet. They are most abundant in the scalp and face and are very numerous around the apertures of the nose and mouth.

Sebum secretion varies greatly with age, state of health, etc. It is subject to the activities of certain endocrine glands and the nervous system. The secretion of sebum is more active during the warm months. It is very slight until puberty, when it greatly increases, then decreasing in the aged. Excessive sebum is termed **seborrhoea**. Seborrhoea is often accompanied by an enlargement of the gland and a number of skin disorders.

Other skin conditions found

Acne: Acne is a chronic inflammatory disorder of the skin. Common acne is known as **acne vulgaris**. It is the commonest of all skin disorders. The outbreaks are generally worse in males than females because the main cause is an increase in the production of male hormones. Normally, the sebaceous glands produces the sebum, but with acne, there is an overproduction that often plugs up the follicle. Acne appears in a variety of different types ranging from presence of comedones (*blackheads*) to deep-seated skin conditions.

STRESS AND ACNE

Acne is only one of the disorders that can arise from stress. Generally an acne skin can be brought under control (*not necessarily cleared up!*) in a relatively short period of time, but an acne condition resulting from stress could become a long-term ongoing problem.   (*Whether or not an individual suffers from acne will be determined within their genetic makeup or in the DNA*). But acne caused by stress can make acne appear for as little as a few days, or as long as years.

Stress does not cause acne in everyone. However, when stress does arise, whether or not it is positive (*i.e.; marriage*) or negative, there will be hormonal changes within the body. The sympathetic nervous system (*under the autonomic system*) begins to take charge in stressful situations and by doing so begins to increase the body's metabolic rate, the heart rate and constricts the intestinal blood vessels (*which can result in poor digestion*) as well as other internal changes.   In response to these changes, the body's adrenal glands secrete hormones that causes the sebaceous glands to produce more oils, resulting in acne. The same is true with premenstrual acne. There is no quick fix under these situations, but realizing the cause and alleviating some of the stress may greatly improve the condition.

Allergic Reaction Breakdown

Here are the steps required to produce a contact allergic reaction.

STEP 1 the inception phase

When you touch a poison ivy leaf, the chemicals from the leaf begin combining with the proteins o your skin at the pint of contact. This stimulates Langerhan cells. An antigen--cell complex is formed and reacts with the T-lymphocytes. Now the T-lymphocytes are activated and they differentiate into new cells with a specific weapon against the poison ivy chemical. These new T-lymphocytes are called suppressor lymphocytes and you have become sensitized.

STEP 2 the elicitation phase

Now that you have suppressor cells in your blood steam, they they circulate around looking to make sure there are no problems. On comes the poison ivy chemical, again binding with the Langerhans cells, but this time the suppressor cells spring into action. Two to four days after contact you begin to itch, get erythema and blister at the point of contact. Anything that binds with proteins will produce an allergic response.

Understanding Sight and Hearing

Sight

The eye is a globular structure filled with a jelly-like fluid under slight pressure to give it firmness. Our sense of sight is the response of the brain to light stimuli and these are received through the eye. The eyeball itself is a hollow, spherical structure, its walls consisting of three principal layers:

1. The sclera or tough fibrous, opaque coat, the sclera being modified in front to form the clear, transparent cornea. In other words, the white outer surface, the "sclera" surrounds the eyeball except for the transparent cornea in the front.
2. The choroid or middle coat which consists of an interlacement of blood vessels and pigment granules supported by loose connective tissue; the iris being a pigmented, muscular curtain suspended behind the cornea. In the centre of this iris is an aperture known as the pupil through which light reaches the interior of the eye.
3. The retina forms the delicate inner layer of the eyeball. In this layer are found the receptor and sensory optic nerve endings sometimes referred to as rods and cones.

DIAGRAM OF THE EYEBALL

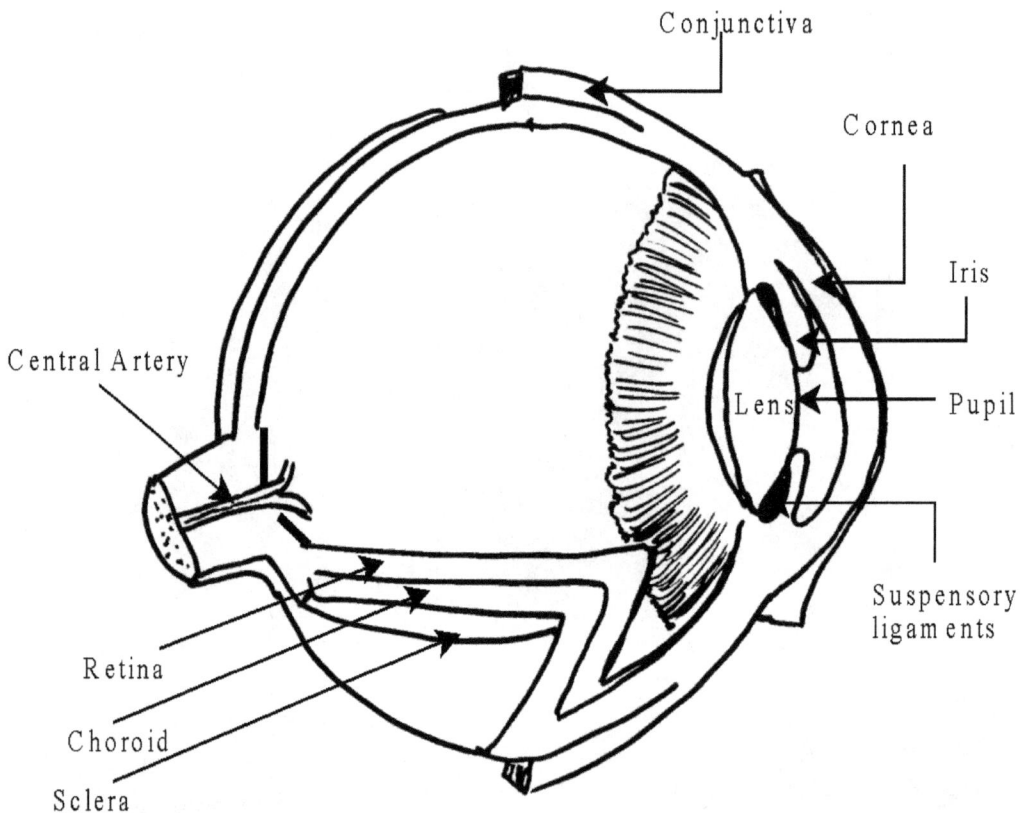

136

The eyeball has a number of appendages, primarily the various muscles which directionally rotate it and the lachrymal or tear glands which moisten and clean the outer surface of the eye. Excess secretion of the lachrymal glands overflow onto the cheeks as tears.

From the inner corners of the eyes the tears drain into a channel which opens into the nose, which is why weeping is sometimes accompanied by sniffling.

The pupil controls the light image by contracting in bright light and dilating in dim light. These light images strike the retina as an upside down image which is then conveyed to the brain through the optic nerve. The brain then re-inverts the impulse so that it becomes a right side up image.

The pupil is a circular opening in front of the lens, varies in size very rapidly depending on the amount of light falling on the retina. The circular and radial muscle fibres in the iris are under the control of the autonomic nervous system. This prevents over-stimulation of the retina by brilliant light. The pupil can vary in size from 1-8 millimetres.

Iris
The iris is the continuation of the choroid in front of the lens. Its colour, genetically determined, depends on the way in which the pigments are distributed.

The lens
The lens is a soft, bi-convex, transparent structure in a thin, tough capsule. It divides the anterior third of the eye from the posterior two-thirds, being held by the suspensory ligaments to the ciliary muscle fibres.

The ciliary body contains the ciliary muscles, which alter the shape of the lens. It is close to the ducts that change the aqueous humour and together with the iris and the choroid coat, forms the uveal tract.

The conjunctiva
The conjunctiva covers the outer surface of the cornea as it bulges forward. Much of the focusing of the eye is made by the convex shape of the cornea, finer adjustments are made by the contraction of the ciliary muscle altering the lens shape.

The aqueous humour
The "aqueous humour" is produced by the ciliary body and circulates through the posterior, behind the iris, and anterior chambers of the eye to bathe the inner surface of the cornea and lens.

Vitreous humour
The vitreous humour fills the posterior two-thirds of the eyeball. Through the centre there is a thin, vessel structure that is the empty remnant of the foetal blood vessel that

used to supply the lens in the foetus, known as the hyaloid canal.

Choroid coat
The choroid coat lines the inner surface of the sclera and has brown pigmented cells to absorb light.

Optic nerve
The optic nerve, containing one million nerve fibres, penetrates the sclera and choroid coats and the nerves spread round the inner surface of the eyeball to form the retina. The point at which the optic nerve enters is known as the "blind spot", as there are no light-sensitive nerve cells at this point. The optic nerve is accompanied by an artery and vein that spread over the retina.

Retina
The retina consists of light-sensitive cells or cones for red, green and blue, and the rods for shades from grey to white.

Hearing

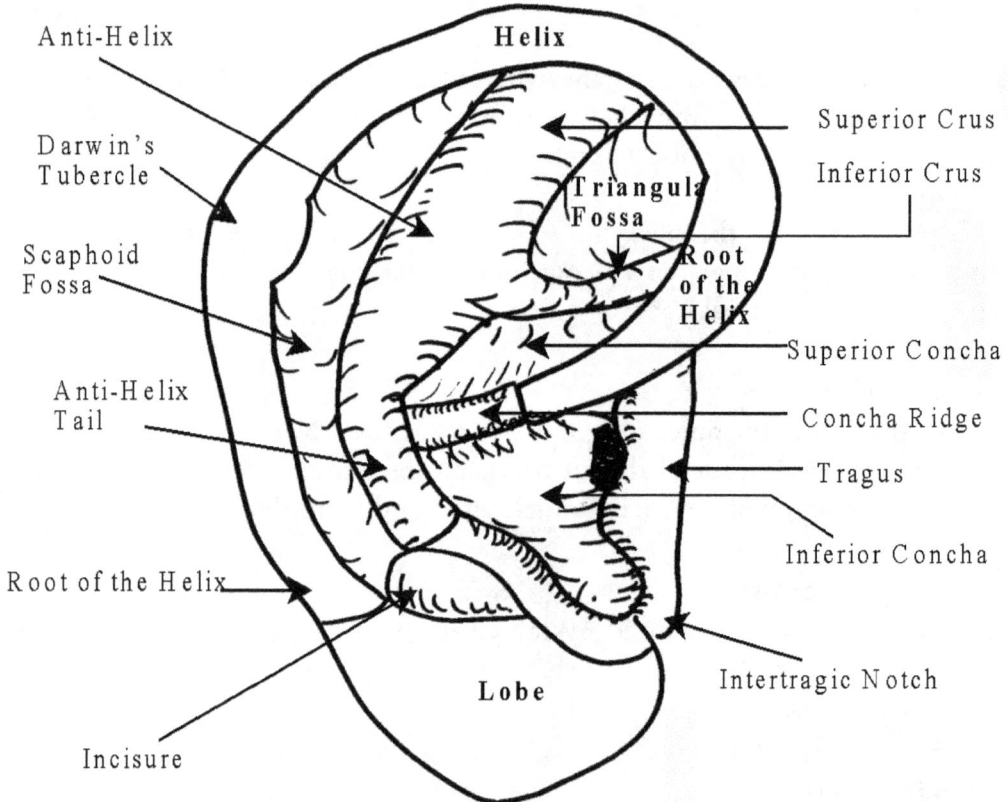

The Ears

The ear is normally divided into three structures:

1. The **external** ear which consists of the **auricle** and the **external acoustic** sometimes called the **auditory meatus**. It also contains ceruminous glands which secrete **cerumen** or **wax**. In between the external ear and the middle ear is the **drum** or **tympanic membrane** which is a parchment-like membrane which lies obliquely across the external and middle ear.

2. The **middle ear** or **tympanum** is a small cavity about five eighths of an inch long by half an inch high by half an inch wide, two principal structures communicate with the middle ear.

   a) The **auditory (eustachian tube)** about one and half inches long, which passes from the nasal pharynx to the middle ear allowing passage of air from the throat to the ear, enabling air pressure on both sides of the drum to be equalised.

   b) The **mastoid antrum** which is an air-filled cavity above and behind the tympanum with which it communicates.

3. The **internal ear**. This consists of three parts:

   a) The **osseous** labyrinth.

   b) The **membraneous** labyrinth.

   c) The **peril lymph** which lies between the two.

   It is the osseous labyrinth which contains the three semicircular canals which are concerned with the control of balance.

DIAGRAM OF A SECTION THROUGH THE EAR

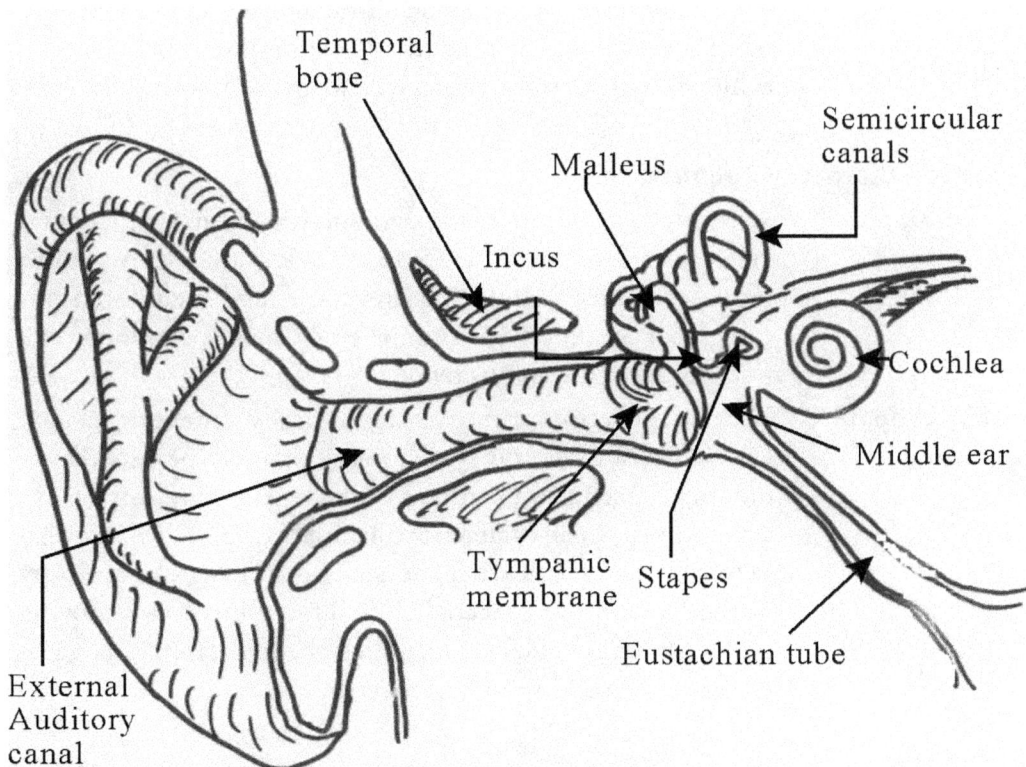

Temporal bone

Semicircular canals

Malleus

Incus

Cochlea

Middle ear

Tympanic membrane

Stapes

Eustachian tube

External Auditory canal

The external ear collects and funnels sound waves or air vibrations to the tympanic membrane. When the sound waves strike the tympanic membrane it beings to vibrate.

These vibrations cause three tiny bones in the middle ear (**the auditory ossicles**) to vibrate as well. The auditory ossicles are popularly called the **hammer**, **anvil** and **stirrup** (malleus, incus & stapes). The vibrations continue to the inner ear, to the cochlea which is filled with fluid and tiny nerve endings. These terminate in the auditory nerve which leads to the brain.

Some Conditions of the Accessory Organs

Disease or Disorder	Cataract
Description	The lens of the eye is situated directly behind the pupil and in health is clear. With age and in some diseases such as diabetes it loses its transparency and becomes more and more opaque, gradually shutting out vision, this condition is known as cataract. It is not a growth but a biochemical change in the lens.
Possible Allopathic Treatment	For the allergic type, cool water constricts capillaries, and artificial tears can sometimes give relieve of symptoms in mild cases. In more severe cases, Non-steroidal anti-inflammatory medications or Anti-histamines may be prescribed. Some patients with persistent allergic conjunctivitis may also require topical steroid drops. Bacterial conjunctivitis is usually treated with Antibiotic eye drops or ointments. Although your GP has no cure for viral conjunctivitis, symptomatic relief may be achieved with warm compresses and artificial tears.
Disease or Disorder	**Conjunctivitis**
Description	This is inflammation of the conjunctiva. In its acute contagious form it is known as 'pink eye'. It is caused by various forms of bacterial and virus infections and includes swimming pool conjunctivitis and the type which is developed as a result of exposure to ultra-violet rays.
Possible Allopathic Treatment	The most effective treatment is to surgically remove the cloudy lens. There are two types of surgery that can be used to remove cataracts: extra-capsular cataract extraction, or ECCE, and intra-capsular cataract extraction, or ICCE. ECCE surgery consists of removing the lens but leaving the majority of the lens capsule intact. ICCE surgery involves removing the entire lens of the eye, including the lens capsule, but it is rarely performed in modern practice.

Disease or Disorder	**Ear Infections**
Description	There are many different types of ear infections including: Otitis Externa (swimmer's ear). The cause is trapped water in the canal, irritating the lining. The moist, dark environment is a perfect breeding ground of bacteria. Serous Otitis media An accumulation of fluid in the middle ear space that is not draining properly through the eustachian tube. It is generally caused by colds, allergies, etc. If the fluid remains in the ear fr an extended period of time, the condition will become chronic and the fluid becomes thick. Bacteria can develop and they may experience hearing loss. Otitis Media A term for middle ear infections. They are caused by either bactera or a virus.
Possible Allopathic Treatment	Antibiotics are commonly prescribed. However, 1/3 of all ear infections are caused by a virus and will not be helped with this treatment as antibiotics are useul in the cases of bacterial infections.

INDEX

www.ingramcontent.com/pod-product-compliance
Lightning Source LLC
Chambersburg PA
CBHW051218200326
41519CB00025B/7162